EXPERIMENTS

NEVER FAIL

Praise for the Japanese version of the prior edition, published under a title that translates to *Are You Happy In Your Job?*

"Business Books That Will Still Be On Your Bookshelf in 10 Years." The book was included in the list developed by Kazuyo Katsuma, one of Japan's leading economic critics and writers, who recommended the book in her *How to Be 10 Times More Effective.*

Nobel Prize winner Shinya Yamanaka (awarded for his work in stem cell research in 2012) says: **"I have read *Are You Happy In Your Job?* many times.** I have to have fun with my job. When I have to decide to do something or not, I'd rather regret trying than regret not having tried. I strongly agreed with the message that *Are You Happy In Your Job?* sends out. When I experiment with something new, it motivates me to try hard, but if I stop experimenting there would be no progress from that point. I always try to experiment."

CEO Takashito Kashino tells *Nikkei Business Online*: **"IMJ Group's corporate motto was derived from a book called *Are You Happy In Your Job?*** People say that only ever-changing companies survive. However, people are afraid of change. In *Are You Happy In Your Job?* it is put as 'People hate to change, but they like to experiment.' It talks about a concept that people can enjoy experimenting as long as they have the right mindset and environment where they can have fun trying. I really liked this idea and decided to adopt it as our corporate slogan: 'Experiment, experiment, experiment!'"

In *Software Design Magazine*, the CTO of Gree, a social networking service and mobile game provider, CTO Masaki Fujimoto listed *Are You Happy In Your Job?* among his favorite books for a feature called "Books Top Engineers Recommend."

Junji Ueda, Chairman of FamilyMart, one of Japan's leading convenience store chains, gave *Nikkei Business Magazine* a summary of *Are You Happy In Your Job?*: "Experiments never fail. People who have succeeded are the ones who changed their goals at some point. In our society, when it comes to time and progress, we tend to view in a linear way, but life is not as punctual as we think. A goal today is a rut tomorrow and it is important to be different tomorrow from yourself today. **Find joy in experiments. Have fun experimenting and be playful when you try. *Are You Happy In Your Job?* made me realize, once again, how important these things are.**"

At a conference on the work of Peter Drucker, Tetsuya Osawa, CEO of Nihon Shokken (a premier Japanese food company), was asked what books he recommends. He listed *Are You Happy In Your Job?* and said of it: "**There are lots of books written about rules of success, but people rarely succeed even when they follow those rules. However, if I can point out one common rule that all successful people share, it's written in *Are You Happy In Your Job?*"**

EXPERIMENTS NEVER FAIL

A Guide for the Bored, Unappreciated and Underpaid

Dale Dauten

MAURICE BASSETT

EXPERIMENTS NEVER FAIL: A Guide for the
Bored, Unappreciated and Underpaid

Maurice Bassett
P.O. Box 839
Anna Maria, FL 34216

Contact the publisher:
MauriceBassett@gmail.com

Contact the author:
ddauten@gmail.com
www.dauten.com

Illustrations by Ted Goff
Cover design by David Michael Moore
Editing and interior layout by Michael Pastore

ISBN: 978-1-60025-201-3

Library of Congress Control Number: 2022948160

For Hilary, Trevor and Joel

Every child is born an artist.
The problem is how to remain an artist.
~ Pablo Picasso

You can't top pigs with pigs.

~ Walt Disney
(objecting to plans for a sequel
to his *Three Little Pigs* cartoon)

EXPERIMENTS NEVER FAIL

Contents

Foreword

Innovative, ingenious and creative.

"I don't read many business books. I read good fiction. Business is about people, so my favorite business books are anything by Dickens."
~ Tom Peters

The great business guru and best-selling author Tom Peters has written many powerful and ground-breaking business books. He's also given a ton of impactful seminars, many of which I used to listen to over and over. Whenever he recommended the business leaders in his audience to read fiction, I was startled. Wait ...why?

But after I'd put in a couple of decades coaching and training leaders I finally saw his point. In the quote up above he says it's because business is about people, and great novels often dive deeper into the nuances of psychology and personal behavior than textbooks and business books have the creative freedom to do.

But I also began to see another benefit to reading fiction. It opens the mind. It lights up the imagination. It reawakens the divine creativity that is our birthright so often veiled in adult life by compulsive worry and problem-solving. We get trapped in circular, ruminative thinking, especially when it comes to business and money.

What's so powerful about this book by Dale Dauten is that it's not only highly readable fiction, but it also contains a huge liberating message for anyone and everyone who owns or works in a business. Be fearlessly creative. Have it be your conscious daily purpose to experiment and innovate. Don't stay trapped in the left side of your brain. Stay open to possibility.

Ever since I first discovered the writings of and then met Dale Dauten he has personified fearless creativity to me. He reminds me of Bob Dylan the day Dylan shocked his folk music following by plugging in his guitar and pouring his kaleidoscopic poetic folk lyrics into the world of rock and roll. A mind-blowing "experiment" that shocked and delighted the world of music.

Dale's work as a consultant to businesses has been a disruptive antidote to corporate bureaucracy and mediocrity. He patiently takes leaders by the hand and leads them out of their worried minds, into the rarified fields of unpredictable imagination. He conducts what he calls "Innovators' Labs" for some of the top companies and organizations (including NASA!) in the world. He frees them up. They learn to experiment. They learn to leave the double bind of "the right thing to do versus the wrong thing to do" into the divine unknown where all innovation is born.

The business leaders I've given his books to are always in for a surprise. They expect just another business book, but from Dale they get things like this: "In EVERY company people are going to make fun of the boss; it's just that in the good companies it happens when the boss is around."

Dale Dauten is one of those talented thinkers and writers who could have succeeded at screenwriting, songwriting, poetry, mystery novels or any of those fields where unusual creativity is celebrated and welcomed. The fact that he chose to bring that talent to the world of business and organizational structures is remarkably bold and ... as you'll experience when you read this book ... beautiful.

Here's how he describes the world his consulting wanders into: "Most jobs are boring because they are designed that way. If you're building an organization, you want to create jobs that qualified people can do readily. Then, when you go to hire people, you look for employees who have successfully done that exact job. In other words, you minimize uncertainty, which is same thing as structural boredom."

Wow. Why would you want to play in that sandbox?

Somehow Dale does, and he's found his way in through the effectiveness of his work. Maybe it was a matter of going where you're most needed versus just going where you're wanted.

When I first read this story you're about to read, I put the words "Experiments never fail" up on my whiteboard in my home office as I was struggling to build my own business. Somehow it would inspire me to try things I might not have tried and not be attached to the outcome. Just to find out what worked and what didn't, and not get seduced by the superstition that shows up in my mind as "fear of failure."

Over time I developed the ability to try new things just to see what would happen. In the creative world of experimentation there is no success or failure, there is just relaxed observation. That habit of experimentation without needing an outcome has served me well over time. And it all came from this book and the story it tells.

Steve Chandler
August 2022

Preface

It's been many years since I started putting the principles in this book onto paper. Many of the ideas went into a book entitled *The Max Strategy: How A Businessman Got Stuck At An Airport And Learned To Make His Career Take Off*. I've always detested that title — it was a compromise with the publishing company that I came to regret.

Now, with more experience and less compromise, and with the inspiration of new publisher Maurice Bassett, I have rewritten the book, updating it and adding a new creativity tool called "Instant Brains."

If you find yourself bored, unappreciated or underpaid, I hope these ideas will be of use to you. Please let me know, at ddauten@gmail.com.

Dale Dauten

"He was having fun; I was miserable..."

Chapter 1

Funny how bad luck has a way of turning into good luck, and vice versa. I've gotten to the point where I don't much believe in good or bad luck anymore, just coincidence.

Take, for instance, a night that started out as one of the worst of my life. I was trying to fly home from Chicago and a snowstorm had shut down O'Hare Airport. It was May, for heaven's sake, but there was a blizzard outside. Not a big blizzard, not by Chicago standards, but the plows had been taken away to summer storage, so the runways couldn't be cleared till morning. (As I write that explanation it doesn't sound plausible, but I swear that's what they told us.)

Whatever the reason they closed the airport, I was stuck for twenty-six hours in one of the terminals of good old O'Hare. At the time it seemed anything but providential. I'd been in Chicago for three days on business and I'd skipped the last meetings so I could catch an early afternoon flight and be home in time to go out to dinner with my wife and daughter.

Instead, there I was in the airport, surrounded by thousands of crabby businesspeople and noisy families. Rather than dining at a restaurant with my family, I was sitting on the floor in my suit, leaning against my carry-on bag, forcing myself to go slow on a box of Raisinets — the last item left at the news shop candy rack.

As I ate the candy, thinking sour thoughts, I watched an old guy clowning around with some children. You could tell that these kids weren't traveling with him — he'd apparently appointed himself Recreation Director for half the preschoolers in the terminal. He was pushing seventy years old, a big, ambling character in plaid pants and a polo shirt and a bolo tie. Given my dark mood, it struck me that instead of giving children rides in an airport wheelchair, the old goat should be quietly thinking of the afterlife. But there he was, bumping around with them. He found everything they did hysterical, and when he laughed, you could see every tooth in his head.

I wished he'd take his kindergarten down the hall so I could sulk in peace. (That feeling was typical for me back then: He was having fun; I was miserable; somehow I was able to feel superior to him.) Eventually the old boy wore himself down. So where does he go? Right toward me, just like a cat who singles out the one cat hater in the room. He lowered himself down against the wall, mopping off sweat with a handkerchief and panting out, "Hello, hello, my friend." Then, without waiting for a reply, he started in on the life stories of his new little pals — one had six brothers, one was wearing a cast on a broken wrist, and on and on.

He settled down and settled in. Then he moved on to some corny jokes. The only one I remember went, "What do you get when you cross a Unitarian and a Seventh-Day Adventist? Someone who knocks on your door for no reason."

Are You Happy in Your Job?

Next, he started asking me questions — nosy questions — all about my wife and daughter and my hometown. Then he asked about my job: What? Where? How?

I answered several of his questions then deflected others with wisecracks. I soon learned that he had more questions than I had wisecracks. Finally he hit me with this zinger: "Are you happy in your job?"

By then it was getting late and my weariness and animosity overtook my politeness and I gave him a true, bitter answer. Before I knew what I was doing, out came all the bile I'd tried to ignore, all the disappointments that had collected on my spirit like dust.

"I'm thirty-seven years old," I began. "I've been working for nearly fifteen years. What have I got to show for those fifteen years? What can I say I've accomplished? All I can say is this: 'I have a good benefit package.' What kind of a deal is that? I'm a loyal, hardworking guy. I've always done the right things, the smart things. I've kept my part of the bargain. But my career has been in slow motion. If I complain about it, all I hear is, 'At least you have a job.' Hey, I'm supposed to shut up and be grateful? That's like saying life is nothing more than not having died yet."

I stopped and looked hard at the old man, wondering if he was wishing he'd never put down next to me. For that moment, I didn't care.

But his expression betrayed no disappointment. Instead, he gave off a professorial curiosity and pumped his grey head a bit, motioning for me to continue.

"The people I work with are great," I said. "The problem isn't the people. The problem isn't even the work. My work isn't hard. It's just work. Work-work. I put in forty or fifty hours a week, week after week, and, whoosh, another year is gone. You get a little raise. The boss tells you that he wishes it were more. He shrugs and tells you that one day the budgets will finally loosen up. He doesn't believe it, not anymore.

"So, like most everybody these days, I'm trapped. Things could be a lot worse, I know. But I can't see things getting a lot better. We all know that the high noon of the American industry has come and gone. What will probably be remembered as The Best Job Market of Our Lives came and went in the first quarter-century of the 2000s and somehow I didn't get rich or famous ... or, come to think of it, happy.   Now, we keep hearing the horror stories of company reorganizations. Everyone has friends who are out of work.

"The upshot is that I'm bored and I'm scared at the same time. Every few months I start thinking that I can't just let my life count down like this, like peeling bills off a roll. I need to take some chances, be bold and brave, and live my dreams. Every few months I buy the latest self-help book and try to get a new dream. I'll set some goals and start saving to open a business. But I don't get far, not with an overstressed spouse who doesn't want to take any more chances, and with two kids and a mortgage and all my other obligations. So I can summarize my life in three words: bored, unappreciated and underpaid."

The Breakout

In the pause that followed, I considered whether or not to tell the old fellow about the time I actually opened a company of my own. It's a big joke to my friends. They remember all my enthusiasm during the planning stage and they love to rub my nose in all the childish expectations I'd had. Usually I avoided the subject, but since I was opening up, I decided to go ahead.

"I tried to break free one time. Three of us got together enough money to open a little business. It was a website development company. I was one of three investors. One of the others ran the thing. Our plan was that after the business took off, we would start expanding, and we'd all work there full-time."

"So how did it go?" he asked, still intent.

"Well, we lost the original investment. Then, refusing to quit, each of us put in more money. I lost my savings. I lost two friends. I lost my dream. When it was finally over, guess what everyone told me? 'At least you still have your job.'"

The Future

"Maybe in a few years I'll have the stomach to try again," I added half-heartedly. "Then again, I don't know if I'll ever have the money to try again. I've had to put money aside in case I get laid off — a new rumor goes around every few months. And every time I see one of those stockbroker's ads about what it costs to put a kid through college, I feel like a schmo for not having started already. And after that there's the retirement planning. Whoa! Retirement. I suppose that's when I can finally stop saying, 'At least I have a job.' Instead, I can say, 'At least I'm not dead yet.'"

Finally I shut myself off. I'd gone too far. The old fellow was probably retired himself and all he'd done was try to be friendly. He didn't deserve to have my problems thrown at him. So I apologized.

He was still focused on me like a botanist confronted with a new plant species, and waved off my apology, saying, "What are strangers for?" Which sounds like a sarcastic remark, but it wasn't, not coming from him. There was a kindness radiating off the man that made me regret that I'd been so sour.

Then, just as he was preparing to comment, a gang of children spotted him. They had found another blasted wheelchair.

Protesting but laughing, he let them pull him to his feet. He turned back to me and put a long hand on my shoulder and let a bit of his joie de vivre tingle into me.

Then he danced off with seven or eight children whose ages, added together, probably wouldn't equal his. I tried to return to my reading, but I was too stirred up.

The Second Stranger

A few minutes later another stranger showed up. This time it was a young woman. She knelt in front of me and pointed toward the careening wheelchair. "Are you with him?" she asked, glowing.

"No, no," I replied, chuckling a bit to let her know that I was too sophisticated to be traveling with a guy in a bolo tie and plaid pants.

"Oh. Too bad," she said, deeply disappointed. I guess she could see that I was confused. She asked, "You mean you don't know who he is?"

It turned out that this strange bird had made several fortunes as an inventor and entrepreneur. She told me his name: Max Elmore. Though I'd never heard of him, I learned that industrialists and politicians thought of him as a friend.

Imagine how small I felt. Here I'd had my chance to soak up a bit of his wisdom and all I'd done was carp about my career. I'd never even introduced myself. Never got the business card out, much less the résumé I kept in my briefcase. And now, if he returned — and why would he? — I could never undo the disastrous impression I'd made. What rotten luck, I thought.

Chapter 2

The old fellow did return, about an hour later. As before, he put his back against the wall and slid down, panting, hamming it up.

I'd been such a jerk during our first encounter that I decided it would be pointless to fawn all over him if I got a second chance at a conversation. I'd just try to get him talking and learn what I could.

The Same Damn Thing

"I've been thinking about you," he said with a look I couldn't decipher. "What you just did was to state the impact of economic shifts on the individual. I've never heard it done better."

Well, I thought, at least I get credit for something: If I'm going to whine, at least I'm a world-class whiner.

He waved good-bye to one of the kids as he continued: "While those little ones were trying to snap my ancient bones, I was turning over in my mind what you said. Two things came to me. One was a quote, I believe it was Edna St. Vincent Millay: 'It is not true that life is one damn thing after another — it's one damn thing over and over.'"

He let out that horsey laugh of his. "And then," he added, still laughing, "the strangest word popped into my mind. It's a word I didn't hear for a decade or two but started coming up again recently. You'll never guess."

"Groovy?"

"Good!" he barked. "That's funny, but it wasn't groovy. The word was stagflation. You know it?"

I did. Great: I had managed to make such a great impression that I had become the personification of simultaneous "stagnation" and "inflation."

He said, "I thought of that word because until the time it was coined, economists had always said that inflation and economic stagnation could not co-exist — you could have one or the other, but not both. And you made me see that our economy has created another impossible pair, this time at the employee level. This time the pair is 'boredom' and 'fear.' You wouldn't think those two would go together, but there they are. People have jobs they don't want, and they live in fear of losing them."

We talked a bit longer and eventually he called it career stagflation. It used to be that as an employee's responsibilities increased, his or her rewards did too. But these days the demands on employees are inflated while compensation in real terms — accounting for economic inflation — is stagnant.

The Search for a Rewarding Career

Although I tried to move the conversation around to him, he kept firing questions at me. He was particularly interested in what I'd done to break out of my rut and find a rewarding career.

He asked if he could borrow the legal pad sitting by my briefcase, then took a plastic pen out of his coat, getting ready to take notes. Believe me, I was impressed that a friend of presidents would take such an interest in me.

The Self-Help Wisdom of the 21st Century

"Tell me your strategy for success — your advancement philosophy," he insisted.

Even though I'd read dozens of self-help books, I wasn't too sure I had a strategy, much less a philosophy. So I talked about the books I'd admired. He put these ideas into two classifications, summarizing in one page the self-help wisdom of the last decade or two.

The first major section was "Goal Setting." Here we

listed a number of the clichés of the conventional wisdom: "You can't arrive if you don't know your destination. Keep your eyes on the horizon. You get from life what you expect from life. Take charge of your life."

The second section was on "Changing Attitudes." I felt a little pride for suggesting the heading, "Build a New You." Here we had items on getting rid of negative beliefs, what I had heard called "hardening of the categories." Then we had some vague notions about getting rid of limiting beliefs and replacing them with positive messages of success, power, energy, and hard work. I suggested this final item: "Don't reinvent the wheel — model your success on success."

After I'd finished talking and he'd finished writing, the old boy turned the legal pad my way and asked, "Do we have it? Is this a good summary of the best thinking on achievement?" I told him it was.

"I want you to post this where you can see it," he instructed. "Put it on the bathroom mirror or the refrigerator or next to your desk. But before I give this to you, I want to do one thing."

That's when he took his pen and made a giant X across the page, right through what he'd written.

Chapter 3

After he handed me the page with the work he'd just crossed out, he took a second sheet of paper and wrote a few words in big letters. He tore off the second sheet, handed it to me and said, "Hang this next to the first page, okay?"

Then he pushed himself up and, calling after an airline employee, strode away. Here's what that second page looked like …

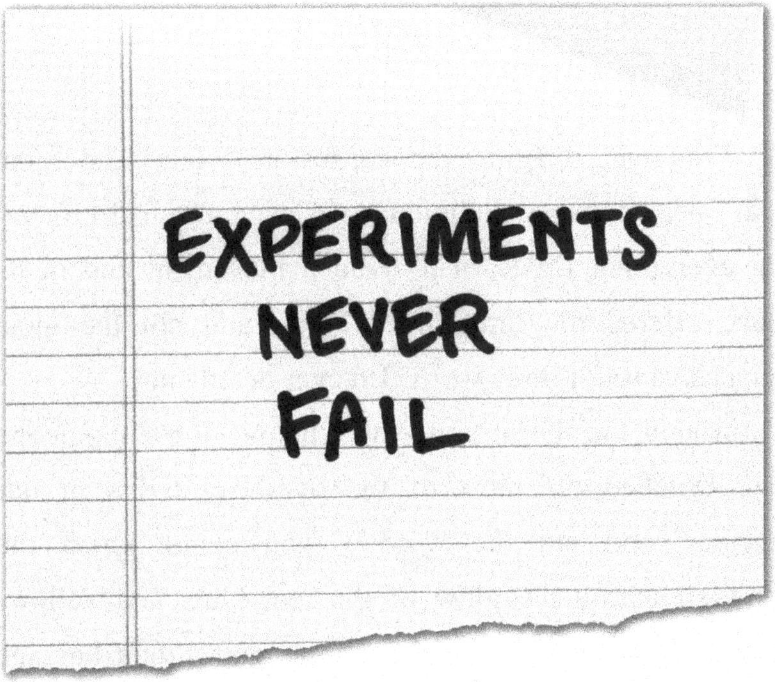
EXPERIMENTS
NEVER
FAIL

Chapter 4

You can imagine the confusion I felt. The most accomplished individual I'd ever met had crossed out everything I'd learned about achievement and in its place offered me three words . . . and not the most persuasive set of three words I'd ever heard either.

Around me the airport was quieting down. Someone had switched off most of the overhead lights in our corridor, and out the window you could watch the snowfall across the glow of the red, blue, and yellow-orange lights outside. People resting against their luggage were beginning to doze off. But not me — I was awake and mentally cataloguing experiments that had failed.

Thank goodness Max eventually returned. If he hadn't, for the rest of my life every time someone had said the words goals or experiment, I'd have been perplexed all over again.

Somehow he had found some hot dogs and Cokes for us. I tried to ask about experimentation, but he was wrapped up in providing me with biographies of the airport employees he'd just met. When I finally maneuvered the conversation back to "Experiments Never Fail," he insisted that we first talk through the contents of the other page, my current "success strategy."

Following his instruction, I got the initial page back out and he reread the first section, about goal setting.

Goal Setting, Revisited

This was his first comment: "You know, I used to go around telling people that passion is the solution to all job problems. It's true, you know. If people are doing work they love, they don't mind putting in long hours, and their zest for solving problems means that creativity flows naturally. With hard work and creativity, everything falls into place. So I give the same advice everyone gives these days: Do what you love!

"It's good advice. But here's the rub: Most people don't know what work they love, what job they'd want to do all day, every day, the rest of their lives. Oh, they might enjoy tennis, say, and would gladly be a world-ranked tennis player, but they know a career as a tennis star is beyond them. So what does that leave?

"Most folks are not quick to admit they don't have passions — that makes them sound like a lump of goo. So they say, 'Well, I'm not sure what my passion is, but I know that I like working with people.'" He sighed theatrically. "The only job that rules out is going to Africa to live among the chimps." After I'd finished laughing, he added, "Or people dance away from the issue by saying, 'I don't know what I want to be when I grow up.'"

Max stared at the ceiling a while and I waited quietly. He sounded melancholy when he continued: "Who can blame the folks who say such things? Most have lived their lives without encountering career passion. Growing up, the only activities where they'd see their parents involved and enthused were sports. Eventually, when a kid realizes that he or she is not going to be a pro athlete, it leaves a big hole. Most of the time that hole never gets filled.

"I've often wondered how different our society would be if parents took the energy they put into kids' sports and instead put it into language or science."

Goal Poisoning and the Silliest Question

He shook his shaggy head, as if he were trying to shake out an idea. "But don't let me digress. My point was that if you don't have an ideal career in mind, you feel incomplete or left out.

"Then again, even if you have a dream of a perfect

career, that's no reason to get too confident. All across the land people are filling up psychologists' couches, moaning, 'I have the job I always wanted but somehow I'm still not happy.' These people are suffering from an overreliance on planning. They have what I call goal poisoning." The energy came back into him and he winked.

Perhaps he sensed that he was losing me, because suddenly he started to rise and suggested that we stroll around the terminal.

"Enough deep thinking," he said. "Let's talk examples. I'm going to throw out the names of some people. Each of them shifted career goals at some point in their lives. Let me tell you what would have happened if each one had committed irreversibly to an earlier dream, had 'taken charge' and 'refused to fail.'"

It was here that Max began grabbing names out of the air and telling me what they might have become if they had stuck with an earlier dream. He told me that legendary basketball coach John Wooden would have become a civil engineer, that the television producer Mark Burnett — *The Apprentice, Survivor* and others — would have used his experience as a former member of the British Army's SAS to become a military advisor in Central America, that writer/poet Maya Angelou wanted to be a stay-at-home wife and mother; and, that as a young woman, Supreme Court Justice Sonia Sotomayor yearned to be a detective.

Then he focused on three people in more detail:

The co-founder of Sony Corporation, Akio Morita, was a first son, destined to take over his family's sake and soy sauce business. However, there was a shortage of science teachers after the second World War and he took a temporary job teaching physics. That's when a friend from the Japanese Navy convinced him to help start a company making converters to allow radios to pick up shortwave signals. (By the way, he mentioned that the original company name was "Tokyo Susuhin Kogyo Kabushiki Kaisha." However, the company's advertising featured a character called Sonny Boy, and that eventually got shortened and became the official name.)

The fashion designer Vera Wang's only goal growing up was to be an Olympic ice skater. When that didn't happen, she studied fashion design and went to work for *Vogue* magazine. That was to be her career, until she got passed over for the editor-in-chief job and quit to take a position with Ralph Lauren. Later, when she was planning her wedding, she couldn't find a wedding dress she liked. That's when her father offered to back her in opening a wedding dress boutique. From there, she became one of the world's premier designers.

And finally, the great choreographer George Balanchine would have become a composer. He did some dancing as part of his schooling, and he was asked to do more and more choreography. Max added that Balanchine's biographer, Bernard Taper, had said of his career decision, "The deed preceded the wish."

To which Max added, "Don't you love that line, The deed preceded the wish. Think of the name "Sony" versus the unwieldy "Toyko Susuhin Kogyo Kabushiki Kaisha" — the advertising character was the deed that preceded the wish for a better name. Think of Vera Wang: The wish was to be an ice skater. Nope. That was a crushing disappointment. The next wish was to be a magazine editor, which went well till it didn't. Another crushing disappointment. The next new wish was to do actual design, not just covering it for a magazine. That worked, till her father made her a better offer and she tried owning a boutique. But notice that she was no ordinary boutique owner — no, she had the Ralph Lauren experience and the *Vogue* experience, which made her unique. And she had a niche: wedding dresses. Did she plan to go into business? No. Did she plan to specialize in wedding dresses? No. She fell into it. Fell. The deed preceded the wish.

"Planning is often little more than wishing, while experimenting is dabbling in deeds. Said another way, experimenting is testing plans before they exist."

Max was interrupted by a woman who couldn't find her sister. Naturally the old boy took over and deputized us to be the search team. Once we had her located, he returned to the goal-setting discussion.

"The silliest question that intelligent people ask is, 'Where do you want to be in five years?' Thank goodness I haven't had a job interview in forty years — I hate that where-do-you-want-to-be question. I plan to be a different person in the future. I don't know who I want to be five years from now, much less where."

Bouncing

These glib pronouncements didn't seem to me to be adding up to anything practical. I stopped him to object: "If you don't have goals, you can't measure your progress. That means you can't do midcourse corrections. You just bounce around."

"Don't knock bouncing," he cried and gave me his barking laugh. "Life is bouncing!"

He grabbed my shoulder and squeezed. "This part isn't going to be easy. Stay with me ..."

The Sole Goal

He continued: "Our society has a linear view of time and progress. Our schools teach it by osmosis — life is a series of assignments, skills to be mastered, grades to be passed. Set a goal and work toward it.

"But life isn't orderly. It gives you the lessons out of order. Life is a schoolteacher's nightmare." He looked at me, asking with a glance if I was following him. I nodded.

He said, "Goal setting makes you feel in control —" then stopped himself. Pointing at the "Success Strategy" sheet still in my hand, he asked to see it.

"Look here — you listed on this piece of paper 'taking charge of your life.' HA! Life is not so tame as that. I've given up trying to tell life what to do; she's not going to listen anyway."

It sounded to me like he was unlocking the secret of accomplishing nothing. I bit my tongue and let him continue. "Most people set goals in order to get out of a rut. Well, my friend, today's goal is tomorrow's rut." Then, pleased with himself, he repeated it more loudly: "Today's goal is tomorrow's rut."

Next he said, "There's only one goal I've ever had. You want to hear it?"

I told him I did.

"To be different tomorrow than I am today."

Chapter 5

He could see I was not convinced. "Maybe you think it's too easy," he suggested. I shrugged, not sure what I thought.

"My one little goal is anything but easy," he continued. "To be different each day? That's the only shot you have at being better. Write this down ...

DIFFERENT ISN'T ALWAYS BETTER,
BUT BETTER IS ALWAYS DIFFERENT.

"You can't get to better without first getting to different. Every blessed day. Believe me, it will wear you out. No, I'm not suggesting the easy way out; this is the exhausting way out. But it's also the exciting way out, the alive way out."

I yearned to object, but he was warming to his subject, preaching. "Life is evolution, and the beauty of evolution is that you never know where you'll end up. Do you think the first fish to crawl onto land had a long-term goal? Maybe that fish thought, 'If I can do this, then I see that someday there will be land fish that walk on legs and eventually the land fish will get around in automobiles and go to shopping malls and walk into the Cinnabon shop and get a roll and a cup of coffee.'"

I had to admit that this Max Elmore had a way of getting a person to think. Even so, he was going too far too fast. This much I could agree with: Every self-help program I'd taken on had come to nothing. I'd examine my life and decide I needed to initiate major changes. At least four times in the past ten years I'd sent away for college catalogues, thinking I'd start over in another field. But somehow nothing ever changed. The goals just slipped away.

Play Around and See What Happens

"Have I persuaded you?" he asked, grinning.

"Maybe," I replied, being generous.

"Not good enough. It might help if we talked about some real achievers. Take the computer industry. Do you know why Steve Wozniak built the first Apple computer?"

"Nope."

"It wasn't because he wanted to change the world or head a major corporation. Back then — it wasn't so long ago, by the way, the mid-seventies — the only home computers were ones assembled from hobby kits. Wozniak was an engineer who built the Apple computer to show off for his pals in the Homebrew Computer Club. No big goals, just showing off. The idea to market the Apple wasn't even Wozniak's; he was talked into it by his friend Steve Jobs. And when Wozniak built the Apple II, it was to prove he could build a computer with a color monitor."

Next he asked me if I knew the story of Bill Gates of Microsoft. He told me, "Gates admits that it's a fluke that he's so rich. He insists that he was never motivated by money. You know what motivates him?" Max barked out, "That computers are 'neat.' That's Gates's word: neat!

"You talk to inventors or innovators and the idea of showing off keeps coming up, along with falling into something interesting. Achievers don't know where they are headed; no, they just figure they'll play around and see what happens. In fact, that's one of my favorite expressions: 'Play around and see what happens.'

"Talk to entrepreneurs and, in my experience, maybe one in ten had a true business plan. Let me think if I have some examples that you could relate to." He thought a while and then poked me with a forefinger and said, "Yes, I've got it:

"First, there's a young woman named Eileen Spitalny,

one of two co-founders of Fairytale Brownies. Why did she start a brownie business? Because her best friend growing up, David Kravetz, called her up and said he was tired of working for P&G and wanted to start a business together, as they had talked about years ago when they were growing up. Okay, but what would the company be? Eileen's mother was known for her brownies, so that became their product, and their 'plan.' The first year they sold two thousand brownies; now it's in the millions.

"Second — and here's a good bounce story — David Anton was working his way through college as a waiter. He got fired. As he put it with a shrug, 'I guess I wasn't a very good waiter.' At the same time, his college fraternity came up with the idea of getting t-shirts printed up for a big football game. The only problem was that nobody had the money to front the shirt purchase. So David used his credit card and got two hundred shirts. They sold out in twenty minutes. So he tried it again … and again. He put himself through school selling shirts.

"When Anton graduated, he took a 'proper' job, selling industrial chemicals. That lasted three months. That's all it took to realize that selling shirts was much more fun. And so he went back to it, and has built it into Anton Sports, a thriving company."

Max abruptly snapped his fingers, then waved one at

me, saying, "Oh, here's another example — Don Winslow, the novelist. He's a marvelous writer. He writes mysteries, my favorite is one called *The Death and Life of Bobby Z.*" He inquired with his eyebrows, saw the lack of recognition in my eyes and rolled his. "Anyway, he's written a string of best-selling novels, but when I met him, I asked about his career, talking the way we are tonight, and he eventually laughed and confessed, 'I am the laziest non-careerist you'll ever meet.'"

"What Winslow meant is that he had no career plan; he was simply skilled at bouncing. After college he had his young heart broken, so decided to move to New York City. Once there, he took a job as an Assistant Manager of a movie theatre. Over the coming months, Winslow began to figure out that something was wrong with the theatre's finances. The numbers weren't adding up. He mentioned his concerns to his supervisor. Two days later he was fired.

"But, the next bounce led to a job working 'undercover' for another theatre chain, which led to a job with a detective agency. Eventually he returned to graduate school. There he took a part-time job portraying a terrorist in a government anti-terrorism program.

"After completing a masters degree, he was accepted into a job with the State Department. As he began to get to know his future colleagues and with them, his own future self, he said, 'I looked at the people and thought, "These will be my best friends."' His response was to take a job with a safari company instead!" Max let out a laugh-roar that fit nicely with the safari motif.

"So Winslow spent five years traveling in Africa and China, and writing when he could. Eventually he grew weary of the travel and joined one of his former professors in a firm doing background legal investigations. In amongst those jobs Winslow produced a series of detective novels, then moved on to his novels about cartels and police corruption.

"He told me he never had a plan, just two interests: travel and writing. And he spoke movingly of his father who used to tell him this: 'When I work, God respects me. When I sing, He loves me.'"

Max was stopped by that one; eventually he intoned the word "breathtaking" with such emotion that I thought for an instant that he might cry or sing, but instead he came to a conclusion: That takes us back to where we began, to bouncing and 'not knowing.' In 'career management,' the failure to have goals is the problem. Then again, in the Zen sense, 'not knowing' is the goal.

"Most people want a career, want to be working toward a goal, and use the 'I don't know what I want to be when I grow up' cliché as a way of pretending that their current mediocrity is somehow temporary. Then again, the same 'grow up' statement can be a rejection of ordinary goal setting. Instead of asking, "What do I want to be?", the question becomes, "How can I keep growing?"

The lesson we can learn from a non-careerist like Don Winslow is that a career does not a life make. Follow your interests and you don't have to know where you are going … you're already there."

Something was troubling me so I cut in, "You say, 'No goals. Play around.' That's fine, but it makes achievement almost a random event."

Instead of arguing with me, he put an arm around my shoulder and gave me such a squeeze that I wanted to yelp. He boomed, "Now you're getting warm!"

Friendly Chance

Max told me that if it hadn't been for the snowstorm that closed the airport, he would have been in the air on his way to London to give a speech on experimentation. Shaking his head, he said, "There are three hundred people who in a few hours are going to show up at an overpriced hotel, expecting to hear my speech. Too bad."

He walked over to some worn luggage, pulled out his briefcase, and returned with what turned out to be his notes for his talk.

"Then again, they'll have so much wine that tomorrow half of them will tell their friends they heard me speak but can't remember a single thing I said. Then they'll conclude that I'm overrated."

He let his eyes travel over the slouching passengers all around us. "Maybe I should give my talk here in the terminal," he said in a stage whisper. From what I'd learned of him, I half-expected that he'd do just that.

Instead, he turned to a passage in his speech where he discussed economist Burton Malkiel's work on stock selection. I'd heard this "random walk" story before, but had never thought of any implications beyond just picking stocks. But Max Elmore had thought of plenty of implications.

"Malkiel dreamed up an imaginary coin-tossing contest," Max explained. "A thousand contestants in a line; heads was a winner, tails a loser. So the thousand people toss their coins and about five hundred get tails and lose. The five hundred with heads toss again. After seven tosses there are just eight coin tossers left."

Handing me the text of his speech, he pointed out Malkiel's description of what happened next: "By this time crowds start to gather to witness the surprising ability of these expert coin tossers. The winners are overwhelmed with adulation. They are celebrated as geniuses in the art

"Given enough chances, chance is your friend."

of coin tossing — their biographies are written and people urgently seek their advice. After all, there were a thousand contestants and only eight could consistently flip heads."

As I read about coin tossers that night in O'Hare, I was only vaguely impressed. It might make for an interesting commentary on investment advisors, but I wasn't ready to accept achievement as dumb luck and I said so. Max just chuckled. "Your problem is that you're logical. Your mind is still stuck in schoolboy mode: Complete all the assignments and you'll get an A. You want a list of assignments."

"Who wouldn't?" I asked, just to test him. "You have me pitch out my goals and what do I get in their place? Coin tossing."

Once again Max gave my shoulder the vise grip. This time he said, "Naturally, you first have to meet all the prerequisites of success. If you aren't smart and hardworking and all that, you're going to fail ten times out of ten. But if you do all the right things, guess what? You fail NINE times out of ten."

He held up his hand before I could object yet again. "Think how many great novels you've read that never became best-sellers. Think how many actors you see in local or regional theaters who are as good as those on Broadway. Their problem isn't talent or work ethic; it's that they aren't expert coin tossers.

"That's why I have just the one goal: to be different, day after day. That means I have to keep experimenting, which is a trial-and-error, bump-around, muddle-through, random, messy business.

"Remember this: The coin tosser who gets the most 'heads' is the one who gets the most tosses. Given enough chances, chance is your friend."

Chapter 6

I t was just after midnight by then and the public address system crackled out the official announcement the runways would not reopen till morning. The weary passengers around us reacted with sighs or grumbles and a few hissed obscenities. It occurred to me then that I was probably the only person in O'Hare to welcome the news. I still hadn't learned what "experiments never fail" was supposed to mean.

Max decided that it was time for me to have a lesson in coincidence by way of the history of innovation. We had, literally, all night for stories and he had an ample supply. His "first career" was as a writer for a business magazine that no longer exists, *The Business of Business*. There, reporting on successes and failures of corporations, their products, services, and advertisements, he said he was "a witness to random genius." The experience started him collecting histories of product discoveries. He began by asking me if I knew the history of what he called "the greatest consumer product of all time." Coca-Cola.

Having once visited the company's museum in Atlanta, I thought I did, but I shook my head "no," eager to see it through Max's eyes.

Coca-Cola

"It's my very favorite product development story," he began. "It happened over a century ago — and, before you ask, NO, I wasn't around to cover the story.

"There was a pharmacist in Atlanta named John Pemberton who was inventing dozens of products for his clientele — remedies with names like Indian Queen Hair Dye, Globe of Flower Cough Syrup, French Wine Coca, and Triplex Liver Pills. Then one day Pemberton goes into the back room of his pharmacy and finds two of his employees — a couple of shop boys — mixing his new headache syrup with water and drinking it . . . and they did not have headaches.

"Pemberton is curious. He tries the stuff. Not bad. Might be better with club soda to give it a little fizz. So he adds club soda and decides to sell it in his shop under the name Coca-Cola.

"And, incidentally, the flowing script logo of the Coca-Cola name was not the work of an ad agency or artist. That's the way Pemberton's business partner wrote the name in the book where the men recorded their accounts."

He paused then and asked me what I thought of the story. I told him that while it was certainly a charming bit of history, I didn't see any relevance for my career.

"Really?" he said, pretending to be startled.

"What should I do," I asked, "bring in some teenagers to hang around my office and hope they come up with something?"

To which he responded, "Not bad. That just might work."

Then he was off to the next story.

Chocolate Chip Cookies

He started by asking, "Some Toll House chocolate chip cookies would taste good right now, wouldn't they?" I nodded. "Do you know how they were invented?" I did not.

"At a hotel in Massachusetts called the Toll House Inn, Ruth Wakefield was making up some of her Butter Drop Do cookies. That day she decided to try a chocolate version. She chopped up a bar of semisweet chocolate. She planned to melt the chocolate and mix it in, but this time she was in a hurry so she took a shortcut and just threw the bits of chocolate into the Butter Drop Do batter, assuming they would melt during baking and give her the effect she wanted.

"The plan didn't work. The chocolate stayed together in lumps. Which, after she tasted the cookies, she decided was a better plan."

"Interesting," I responded, "but I don't know how you could generalize from it, or from the Coca-Cola example either. A fluke is a fluke."

"HA! You are a tough case, kiddo. Yes, a fluke is a fluke. But you could use a fluke in your career, no? So maybe we should learn their secrets and become 'flukologists.'"

If I have described my companion with any success, you can picture his belly laugh as he slapped my knee, having delighted himself with the word he'd just coined.

Then, we returned to the flukes.

Levi's Jeans

"Levi Strauss came to America as a teenager," he said, "and settled in Kentucky, working as a peddler. But when he heard about gold in California, he eagerly signed on for a clipper ship to San Francisco.

"To earn money for miners' supplies, Strauss packed merchandise to sell to his fellow passengers. Great plan. Sold it all. Everything but some lousy rolls of tent canvas that no one wanted. So, when he arrived in San Francisco, he tried again with the tent canvas. Again he failed. But while he was out in the marketplace he learned that the one product that was in short supply was pants. The job of prospecting was tough on trousers.

"So Levi Strauss hired a San Francisco tailor to make overalls out of the canvas. Demand was so great that the young peddler gave up his prospecting plans, having struck gold in San Francisco."

The Father of Invention

Once again Max looked for my reaction. Part of the joy of conversing with him was that you knew a tough question was never far away. This time the question was, "Why do you think I'm telling you these stories?"

I replied confidently, "To prove your point that innovation is a disorderly business, that it's playing around and bumping around."

"You, my young friend, have got it. 'Ideas' and 'accidents' are half-sisters. Once you start paying attention to them, you see coincidences everywhere you look."

Here I hesitated. "Everywhere you look?"

"That's right. I told you about Coke and Levi's and Toll House cookies because those are products you know — some of the great products of all time. Not every idea is a Coca-Cola idea. But the ideas happen at every level, and all the time."

"For instance?"

Max Elmore didn't hesitate. "I once spoke with Stephen Cuomo of Cannondale Bicycles, who told me how the bike got its name.

"The original company was just the founder, Joe Montgomery, and two employees in rented space above an old pickle factory. First day on the job, the owner sent one of his employees to call the phone company to arrange to get a phone line. So the employee went to a pay phone down by the train station. Well, the phone people insisted on knowing the company's name. Makes sense. But the owner hadn't gotten around to picking a name. So the employee decided to make up a temporary one in order to get the process rolling. He looked around and saw the sign for the train station: CANNONDALE STATION. So he said Cannondale. Everyone liked the name and it stuck."

Before I could comment, he said, "No, here's a better example — the China Mist Tea Company. They sell iced tea to restaurants across the United States and Canada. They're famous for their unusual tea flavors.

"Not long ago I asked the co-founder, Dan Schweiker, how they came up with all those flavors. He told me about the Prickly Pear flavor, taken from the fruit of a cactus.

That one was a request of the owner of a restaurant in Santa Fe. Sometimes you don't even have to toss a coin — the idea walks in the door.

"But here's the example I wanted to tell you about — their Mixed Berry flavor. That one came about because in the process of making up other fruit-flavored teas, they ended up with three partial batches. So instead of throwing out the partials, they threw them all together . . . mixed berry. It became one of their most popular teas."

Max summed up by saying, "Businesses and careers are just like everything else in the world: a series of coincidences. Most people agree with the old cliché that 'nothing works out the way you plan,' but most people still revere planning. I say we need fewer planners and more flukologists."

And then he uttered a line that I have since said to myself many times: "Necessity may be the mother of invention, but coincidence is the father."

Chapter 7

Max Elmore then returned to the "success strategy" sheet that he'd created with me earlier. Of course it still had his big X through it. He wanted to return to the section I had called "Build a New You." That's the section where I had argued for replacing limiting ideas with positive thinking and role models.

Between the Cracks of Logic

"Let me tell you about one of the realizations that changed my life," he began. "It happened when a thought came to me that made no sense. That's where real learning is — between the cracks of logic.

"I was a much younger man, about your age, still working as a journalist. I was covering a conference on what they called 'attitude research' — a bunch of intellectuals intellectualizing about consumer purchasing decisions.

"The basic premise was that a consumer goes through

a series of steps — learning about the product, liking the idea of the product, coming to prefer it, and then, at last, making an informed decision to purchase. You develop attitudes and then your behavior follows. Tidy. Logical.

"But it happened that the day before this conference I'd gone to the grocery store. And I started thinking about the items I'd selected. Soap, for instance. I'd walked past the soap displays and saw the package and liked the color and thought, 'What the hell,' and I bought it. It wasn't education, then decision; it was decision, then education. What the hell. Not so tidy. Not logical. Messy. Just like life."

"Of course," I added, trying to be helpful, "you're talking about unimportant purchases. Little items. Impulse purchasing."

Instead of agreeing with me he raised his eyebrows, leaned his head to one side, and asked me to recall the last time I had learned all about competing products and made a careful, informed decision. I said that I had bought my last car that way.

"What was it?" he asked.

"I bought it just to commute to work. An ugly, old Volvo that I got a great deal on. Solid. Practical."

"And if you had it to do over, would you buy it again?"

I grinned, feeling found out. "Not a chance. I'd buy

something I really wanted."

He grinned back. "You needed more what-the-hell and fewer statistics. Facts are bullies. They push out real information. If you want a car you can fall in love with — start with the car, then check the facts. Don't pick a car — let it pick you."

This advice rang true, although I wasn't sure I could explain why. Maybe I had gone "between the cracks of logic."

He continued: "And you can see why I'm not so interested in attitudes anymore. Once I used to ask people, 'What do you think?' Now I ask, 'What have you tried?'"

Max chuckled. "It's funny how people get attached to their attitudes. Remember José Torres, the boxer?" I'd never heard of the guy, but I nodded anyway, knowing a story was coming. "Torres said that before any big fight you could go down to the Bowery and ask a bum who was going to win. Then the next bum would disagree. They'd argue for hours, maybe get in a fight themselves. José Torres said, 'That's the great thing about boxing — one bum is always going to be right.'" Max grinned, then added, "And that bum is going to think he's a genius. Reminds you of the coin tossers, doesn't it?"

"Sure. But I'm not sure I can tie either one to a career."

He responded, "That's where experimentation comes

in. You've got to let go of maps and just explore. You may think that if only you could have the right set of attitudes, you'll get whatever you want. But the best you can hope for is to like what you get.

"Some fellows did a study of sixteen companies that had produced new products that changed the world — microwaves, VCRs, overnight delivery — that level of innovation. They put it in a book called *Breakthroughs! How the Vision and Drive of Innovators in Sixteen Companies Created Commercial Breakthroughs that Swept the World.* Guess what they found?

"They found that how you organize the company doesn't matter. Their conclusion was, 'You can get a breakthrough whether you deserve it or not.' The companies didn't develop the right attitudes then get breakthroughs; they got the breakthroughs and later pontificated about the attitudes.

Imitation Success

My face must have shown the confusion I was feeling. "It will make more sense as we go," he said and grinned. Then he went back to the "Success Strategy" page and reread the parts about finding role models and "modeling" achievement.

"What is written here," he told me, "boils down to this

philosophy: 'Imitate success.' As for careers, the more successful people I meet, the more convinced I am that grand ambition and lofty career goals are NOT correlated to achievement, at least not in the way the self-inflating purveyors of how-to-succeed books contend. Two case studies:

"First, let's look at Roy Vallee, who was the CEO of Avnet (a Fortune 100 computer component company), who has lived the corporate version of the American Dream — starting out stocking shelves, rising to CEO. There's just one thing missing from the old American Dream scenario: the dreaming. Vallee told me that he never set a long-term goal for his career. In fact, it never occurred to him that he might be CEO until he was made a division president. And listen to how he came to that job …

"Vallee was a Regional Sales Manager, one of many. And while he had no long-term goal, he maintained a single short-term goal: to be the best in the company at whatever job he had. Along the way, he was persuaded to take on an additional assignment, working with a manager at Motorola with the goal of improving relations between the two corporations. Vallee helped organize a joint meeting for executives of the two firms, and the CEO of Avnet was persuaded to attend. As Vallee made his presentation, the CEO leaned over to the Director of Marketing for Motorola and whispered, 'I wish we had people like that at Avnet.' The Motorola executive chuckled and replied, 'You do — he's one of yours.'" Max delighted himself delivering that line and waited till I showed I was delighted, too.

"Here's the rest of that story: The next day the CEO called Vallee and asked him to dinner, and offered him a new job, heading the computer division of Avnet. That night, chatting with his wife, Vallee had a revelation: He might just run the company someday.

"Now for the second example. A coach named Lute Olson turned The University of Arizona basketball program into one of the best in the nation, including three trips to the Final Four and a national championship. Yet head coach Lute Olson insists that he'd never planned to be a college coach, much less win a national title. He said, 'My whole focus in college was to teach and coach in high school.'

"That comment caused me to respond to him by saying, 'So you weren't especially ambitious?'

"He acted surprised, saying, 'Oh, I was VERY ambitious: I wanted to build a great high school program. To do that, I figured I had to improve each year, to be a better coach this year than last.' Only after thirteen years of improving as a high school coach, did Olson decide he wanted to move to a junior college. He eventually moved on to a small college, then Iowa, then Arizona, then national fame."

Max raised a finger to signal that he was about to deliver The Conclusion: "What Roy Vallee and Lute Olson have in common is that they didn't have The Big Dream; instead, they had The Useful Dream — to be better."

I told him I didn't want to nag, but I was still having trouble connecting all the philosophy to my situation.

He smiled wearily, and I feared I'd disappointed him. But he rose to the challenge: "Okay, let's assume you are, oh, say, an ambitious corporate employee. Compare for a minute the effect of these two dreams:

1. I want to be company president.
2. I want to be better this year than last.

"The first is likely to make you petty and defensive, a perennial member of the Hindsight Committee, obsessed with your who-gets-credit ratings. But, if you have the goal of being better, you will be forced to take risks, to experiment, to seek help and offer credit.

"The how-to-succeed literature would have us forever looking for successes to imitate — 'role models' are

*The Hindsight Committee,
obsessed with who gets credit.*

just human goals, and have the same shortcomings as all goal-setting."

He waited for my reaction, and my hesitation was enough to prompt him on to another example. "Okay then, let's talk about something more specifically relevant to you. Let's talk about that business you opened and see how it fit your success strategy and mine. Tell me the story again."

Working Out the Bugs

I explained that our business venture happened about a decade earlier, not long after I graduated from college. Three of us were talking about breaking free and starting a business together. We decided that among us we had all the skills for business success. We met on and off for months, doing our homework about what the experts say it takes to make a business succeed.

One of the guys had just come back from L.A. where he'd visited a company where they designed websites for small businesses. It was mostly a big open space with a bunch of desks and young guys creating new sites. Now you have lots of options for website design, but at the time we thought we'd "get in on the ground floor."

We ran the spreadsheets and put together a business plan and opened up. We'd "work out the bugs," then expand till we had a major enterprise, and then we'd sell it off.

One of the guys was out of work, so he was our full-time manager. The other partner and I put up money and helped out part-time.

We ended up getting a little office space in a rundown strip mall, chosen because it was the cheapest lease. We couldn't afford much advertising, but we grew slowly by word of mouth. After about six months we got to break-even, paying the full-time partner a modest salary. We thought we were getting things right, that we were "working out the bugs." Another year of steady growth and we could move ahead with plans to expand.

Then we got competition. A company came into town that had a real marketing effort, including salespeople out calling on businesses, and who had some specialized talent and software that enabled them to work at a higher level than we could. We hadn't gotten that far. It hurt us a little, but nothing like when the do-it-yourself companies started advertising. So on one side we had competitors who were bigger and better, on the other side competitors who were cheaper. What could we do? We put in more money but we couldn't afford to try to match the one competitor on quality and when we tried to compete with other competitors on price, we'd fall below break-even.

So we just limped along until we managed to get out of our lease."

I concluded my recollections by saying, "I spent hundreds of hours and thousands of dollars and it was all gone. So many resentments built up among the three partners that now we hardly see each other. So much for taking a chance and breaking free."

Some Lousy Secret

We sat in silence a while, in honor of the passing of one more American Dream. Max was the first to speak: "I've heard stories like yours a hundred times and every time I feel the pain that the failure caused."

I decided to give him a little dig: "I guess you'd have to say that our business was an experiment that failed."

He grumbled, "You rat," and slapped me playfully on the arm. "We haven't gotten that far yet, but we will. For now, let's see what your experience says about your success strategy.

"First, let's go back for a minute to goals. You had goals. Like most goals, yours were really just wishful thinking. You were counting imaginary money. But you had goals. And you had dreams.

"I've never known a business to fail for lack of goals and dreams. Funny how we hear over and over that dreams and goals are the secret to success, but when people actually implement them and enter the marketplace, they fail nine times out of ten. Some lousy secret, those goals and dreams."

Being Different

"The problem with your goals," he told me, "and everybody else's too, is that the world wouldn't hold still long enough for them to work out. Technology was changing, competition was changing.

"Think about this: What would have happened if your single goal for your business had been the one I'm suggesting for a career — being different each day? Remember that the only way to get to 'better' is to go through 'different.' With that goal you might have expanded into other services, or added new equipment or new marketing techniques. You might have stumbled upon a new niche, or found jobs that didn't exist before; instead, you merely followed the leader and therefore had no chance for discovery.

"But I'm getting off the subject. Not only did you have goals and dreams, it sounds like your attitude was positive and you found a company to serve as a role model. In other words, you met the conditions of your success strategy. According to your formula, you were certain to succeed.

"This is not uncommon. Most businesses start out just the way you fellows did — trying to grab onto someone else's success. But every other entrepreneur is out there doing the same thing. The imitators flock in and are so busy fighting each other that they can't keep up with the leader, the innovator."

Success Without Rules

"And the same thing happens with careers, not just businesses," Max insisted. "You try to imitate role models, but that's just another route to the same old ideas everyone else has. Worse yet, when you talk to the role models, they'll usually have cleaned up their own histories to make them look more professional — which, said another way, is to make them look more conventional — which, said another way, is to make their success stories look more like everyone else's.

"No matter where you look, people are trying to figure out success by looking at success. It reminds me of something the great novelist Somerset Maugham once said: 'There are three rules for writing a novel. Unfortunately, no one knows what they are.'" He laughed, and I did too.

"It's a bit different with careers and business —

everyone knows the rules of success. You can find them listed in a thousand books. But there's still this problem: Studying novels doesn't make you a novelist and studying success won't make you successful. People try to conform to success, but to be successful is to be a nonconformist.

"Let's put it this way: You don't become a Picasso by taking a Picasso print and running it through the copy machine.

"All of which explains why I put that big X through your success strategy. If you wanted to design a strategy for failure, either in your career or in a business, you couldn't do much better than to concentrate on goals and imitation. With that strategy you'll fail nine times out of ten, whether we're talking about a career or a new business."

A Harder Lesson

"Now, are you ready for the even harder lesson?" Max asked. "If you take my advice and innovate instead of imitate, and work every day to be different from yesterday, you'll improve your odds: You no longer will fail nine times out of ten . . . ." He wagged his eyebrows like Groucho Marx, then concluded, "You'll fail eight times out of ten."

Chapter 8

You can imagine how let down I felt. I wondered if our whole conversation wasn't some kind of overwrought gag. I told the old fellow that I would have trouble getting excited about a program that offered failure eight times out of ten.

We had returned to our spot along a hallway. He smiled as he eased himself down to sit on the carpet, then said, "If you think about it, succeeding two times out of ten is not so bad. Don't you wish you could buy lottery tickets with those odds?" I had to agree.

The Achievement Lottery

"Real achievement is a kind of lottery," he continued. "You enter by being competent and hardworking. Most people get one shot in the lottery, playing at one-in-ten odds. I'm trying to show you how you can enter again and again, and at two-in-ten odds."

"Here's the logic. Most people pick one career and

they do what's expected of them and try to be like the successful people in their field. The result is that everyone does what everyone else is doing. If a great new idea comes along, sure, they adopt it. So does everyone else. You see what is happening to each of them? Each is trying to be exceptional, but ends up going about it by being just like everyone else. The upshot? They have, at best, a one-in-ten chance of being in the top ten percent of their profession. One career, one chance, one-in-ten odds."

The New-Life Gamble

"Think about starting a new career or a new business. Most people never do it because it means they have to risk everything. Even if they get decent odds of success, it may be too scary.

"There's a psychologist I interviewed — John Mowen — who taught a bit about the psychology of gambling.

"Say you have a net worth of one hundred thousand dollars — if you sold your house and car and everything else, you'd have a hundred thousand dollars in cash. And say I told you that I'd bet you a million against your hundred grand. We'd each roll a pair of dice, high roll wins. That would be a ten-to-one payout, with a fifty-fifty chance of winning. Would you do it?"

Before I could say yes, he asked me to imagine telling

my wife about my plan, and to think of my daughter and about every possession we own. He said to think of a fifty-fifty chance that we might lose our house and car and furniture and wedding rings and have to start all over with nothing. When I heard him put it that way, I had to say that I'd probably pass up the bet.

"That makes you like most people," he responded. "And that proposition was fifty-fifty, not the more realistic one-in-ten. No wonder so few people venture outside their ruts."

The Continuous Gamble

He let me ponder his gambling scenario for a while, then asked a tricky question: "But what if you had the same offer, but cut the amounts in half? You might be more willing to go along if you didn't have to risk all that you own.

"And here's another one: Even with the original numbers, what if you could play more than once, if you had multiple chances to win? Say you could have ten chances. You'd be crazy to turn me down. Odds are that you'd win five times and lose five times. So you'd win five million dollars and lose five hundred thousand. You'd come out four and a half million ahead."

This sounded better. I made the mental calculation —

even if I lost nine times and won once, I'd end up ahead.

"Well, the real world offers a deal that is partly better, partly worse. As I already said, the odds of achievement aren't fifty-fifty, they're much worse than that. But here's what you should know: In the achievement lottery you get hundreds of chances to win and most of them don't cost much."

The Achievement Game

"Here's where we are so far," he explained. "We want to play a game that has a big payout, and one that doesn't cost too much to enter, and that we can play over and over again and keep winning."

By now I was able to guess that the old fellow was building up to his earlier assertion that "experiments never fail." For one of the few times that evening, I was right.

Stop Being Ordinary

Next he laid out this simple truth: "If you want to be extraordinary, the first and hardest step is to stop being ordinary."

This, I understood: To play the achievement game a person had to work at being different in order to work at being better. Doing that meant continual experimentation.

Next he offered me two rules for becoming an experimenter:

"One, there's never a 'right time,' or a 'perfect opportunity.' Which means that you begin right here, right now.

"Two, the obvious ideas have mostly been used up, as have many of the non-obvious ones. The result is this: If it weren't for long shots, there'd be no shots at all."

Experiments Never Fail

It was, at last, time for Max to explain to me the logic of his statement that experiments never fail.

You'll remember that I had chided him before, telling him that my attempt to start a business had been an experiment that failed. He finally was ready to debate my conclusion.

His logic was that our company had not been "an innovation but an imitation." Even so, he reminded me that the company had begun to prosper, failing only after competition had moved in. "Your plan apparently was okay," he said. "It worked for a while. But you weren't experimenting, weren't getting better. You were a stationary target for your competition. Anyone who wanted to move in could come in and see just what you were doing and then add a few twists and do it better. You were doomed." Then he spoke softly a sentence that is still reverberating in my head: "Your business wasn't an experiment that failed, but a failure to experiment."

He explained: "You and your partners had a vision of the perfect little business for yourselves. Your goal was to open up and then 'perfect' the business. That's always a mistake."

Once again he was starting to lose me and he read the confusion in my face immediately. So he launched into a remarkable story. After hearing it, I never called myself a "perfectionist" again.

Better Than Perfect

Max described having once attended a "master class" in flute taught by the late Jean-Pierre Rampal, regarded as the world's greatest for decades. For the class, the best young flute players in the country came together to play for the master. After each one played, Rampal would take up his golden flute and say, "Perhaps like this" and then replay a passage. Both were beautifully played, but often they were very different.

Finally, Max asked Rampal about his willingness to reinterpret a composer's work. This is what Rampal said: "There are nights that I go out and play a piece perfectly. Then, the next night, I go out and play it better."

Whoa! Better than perfect. That doesn't make logical sense, yet it was clear to me the minute I heard it. That line changed forever the way I think about perfection. Perfect isn't good enough — you still need to experiment.

The old boy added another line I like. He told me that once you decide something is perfect, it sits still, waiting to be overtaken by competitors. The result is that, as he put it, "Perfection is the first stage of obsolescence."

Once I'd heard Jean-Pierre Rampal's quote about "better than perfect," I abandoned my pride in being a "perfectionist." But, I could still name plenty of experiments that failed. And I proceeded to do just that.

Max waited me out, then replied: "I'm not saying that everything you try is going to work, or that every decision will be a good one. Of course not. I've already told you that most plans don't work. What I'm trying to get you to see is that even though most ideas don't work out, experimenting does.

"I guess I could have said, 'Experimentation never fails,' which would be true and easier to understand. But you know what? When you try something and it turns out to be a lousy idea, you never really go back to where you started. You learned something. Maybe it makes you appreciate what you were doing before. Plus — and this is big — you've practiced being different, which is the key skill of achievement. So I think it is true that experiments never fail.

"I want you to believe it too. And that doesn't mean I want you to talk solemnly about 'the scientific method' and 'control groups' and all that rot. Just play around, try new things, and keep your eyes open. The hard part is trying to get people to change, to be different. Most people hate change. But here's one of the most important sentences ever to come into this gray head."

He put his fingertips on his temples, as if to push out the next thought. "People hate to change, but they love to experiment."

Chapter 9

O'Hare Airport had grown still. Someone had switched off part of the lighting, so that in the half-light the corridor looked as though it were lined with boulders — people and their luggage blurred together into gray lumps. Only two people were fully awake, and I was one of them. The other was eagerly guiding me into a new topic of conversation: the Hawthorne Effect.

The Hawthorne Effect

If you took sociology or psychology classes in college, you've probably heard of it. My own recollection was that the name came from an early piece of industrial research. The goal was to see what changes caused factory production to increase or decrease. But, to the researchers' surprise, production went up, regardless of the change.

They concluded that the workers liked being part of the research, so much so that it caused them to be more productive. So the "Hawthorne Effect" came to be associated with research that was muddied by the fact that the participants knew they were part of an experiment.

Max told me that he had written about Hawthorne as a magazine reporter, saying, "I wrote about it decades ago and I haven't stopped talking about it since. Throw out all the industrial psychology but keep Hawthorne and you'll be ahead. Better yet, keep just one book, a report on Hawthorne called *Management and the Worker*, and you'll be even further ahead."

The Hawthorne Works

The Hawthorne Works was a huge telephone factory in Chicago. It was the largest plant of the Western Electric Company, the production unit of AT&T. The experiments began in 1924, and they began without grand aspirations. The original goal was simply to study the effect of lighting on the workers' output. The question was, should the factory have more light?

Here's how Max described the findings: "Simple experiment, simple results. The more light, the better. As lighting increased, so did production."

The First Surprise

"But, then came the surprise: When the experiment was

essentially over and lighting was returned to standard levels, production did not drop off.

"The researchers were confused and so they repeated the experiment with a 'test' and a 'control' group. The test group had three different lighting intensities; the control group remained at just one. That should clear up the findings, right?

"Nope. Production jumped up for both the test and control groups."

The Next Surprise

"You can imagine the response of the researchers. Perplexed, they decided to try again, this time decreasing the light in the work area. As the level of illumination fell, output rose. The decrease continued until the workers began to complain that the shop was so dark they were barely able to see what they were doing. Only then did production slip.

"At last it occurred to the experimenters that something was up, and it wasn't the lights." As he hit that line, he hit me with an elbow; I was so intent on his description that he almost caused me to topple over.

The Not-So-Controlled Test

"Now we get to the good stuff," he said, wagging his eyebrows again. "The researchers talked management into doing a new series of experiments. This time instead of lighting, they would test the effect of 'workbreaks,' or what you or I would call 'coffee breaks.' They picked just five volunteers and moved them into a separate room.

"The research was, by modern standards, sloppy. In today's 'perfect' research, just one variable is changed at a time. Looking at Hawthorne, there were all kinds of variables introduced all at once, including the move into the smaller room with its better lighting and improved workbenches.

"But the volunteers — all of them were women, by the way — were also given incentive pay based on the output of their group of five, instead of on the old, larger group of one hundred. And some rules, like ones about talking, were relaxed. And the workers were the object of plenty of special management attention.

"Into this new environment came what was meant to be the test variable: workbreaks of various lengths.

"The researchers were amazed when this group's output rose thirty percent, regardless of the break schedule used. Most amazing of all was that when the women returned to the standard schedule, with no breaks, output reached a record high. So you could say that the experiment was a research disaster. The goal of the project was to draw conclusions about workbreaks. What conclusion could be reached? On the other hand, output had risen thirty percent! Why?"

Pay Versus "Zest"

If he'd waited for me to guess, I would have said that the difference in incentive pay was the key. But, Max explained, the authors of the study report concluded that differences in pay were small and could account for, at most, half the increase. So what about the other half?

Max pulled out the text of his speech and showed me just what the authors concluded back in 1939 about the new work environment:

"[The volunteers] lost much of their shyness and fear, or what came to be called their 'apprehension of authority.' They talked more freely among themselves and about themselves to officers of the company and to the observers. They developed greater zest for work. New personal relations between members of the group arose which developed into strong bonds of friendship. They visited one another's homes and went to parties, dances, and the theater together . . . At work, it was not uncommon to find one [worker] increasing her output so that her friend, who might not be feeling well, could rest."

The Lessons of Hawthorne

After Max Elmore described the study, he offered his own view of what it all meant. "When you hear about 'the Hawthorne Effect,' it's usually in the form of a warning about the failure to control for the attitude of the research subjects.

"Well, modern sociologists picked up the wrong end of the bat, learned the wrong lesson. They think of Hawthorne as sloppy research, as an experiment that failed. But we know better. The grand lesson of Hawthorne is that experiments never fail. Look at all we can learn from Hawthorne:

"One, people love to experiment. They volunteer!

"Two, people want to be part of a team, and a 'test' group is an elite team. Once people believe they are part of a team, they start to help one another and thus take over much of the job of a supervisor.

"Three, modern researchers keep looking for the 'perfect' research design. So they get narrower and narrower until they see nothing at all. They miss out on the synergy of life. Sneak in one little change and nothing happens. But overtly change these nothings all at once and you get a big something — in this case, a thirty-percent increase in production. The lesson of Hawthorne is to change everything and then change it again."

Something New

Max turned his face toward the ceiling, contemplating. "I suppose I could say that it's the joy of experimenting that keeps me working. And by inviting people to join me in the experiments, I'm part of all sorts of teams. The team motto is always the same: 'Let's try something new.'"

He turned back to me and gave my leg another squeeze as he did one of his barking laughs. "My life," he told me, "has been one long Hawthorne Effect."

Chapter 10

Max stretched out his legs and peered at the dozing travelers. He said softly, "I should let you get some rest."

When people use an expression like, "I should let you rest," you're left to wonder, is that for my sake or theirs? Usually I assume the latter, but not this time. I wasn't tired. I wanted to hear more and said so.

Working Magic

"Really?" he responded, genuinely surprised. "You have a large capacity for difficult information." He stood up and said, "I'm getting stiff; let's take another stroll."

As we passed those sleeping against the walls of the corridor, mostly businessmen in blue or gray suits, my companion turned a bit melancholy: "Look at all these people. They're not much different from the farm workers in England who went to London in the early days of the Industrial Revolution. These poor kittens sleeping here in

O'Hare are caught up in the greatest economic shift since those days and they don't even know it. Most of their employers don't know it, either.

"Look at them, my young friend. Beaten down. Worn out and frustrated. Dreamless. In the eighteenth century there were factories where people were worked to death. Today it's mental work instead of manual work, but our society is just as cruel. The mental burdens of workers in our time will one day be looked back upon in horror."

"What went wrong?" I asked.

"Everything and nothing. Foreign competition. Both employers and employees got more competition. In other words, we've been slapped around by the 'invisible hand' of capitalism." He clapped his hands together, startling me for an instant in the still corridor.

The Big Squeeze

"We live in a time of excess capacity, not just in the old sense of industrial capacity — too much aluminum or too many semiconductors — but too many businesses of every ilk. Sure, there can be international events that cause supply issues, or we get hit with economic shifts, but in the modern society, those are always temporary. We have more fast-food restaurants than we need, more dry cleaners, more print shops. And there are too many professionals — too many accountants, too many lawyers, too many writers.

"And that oversupply brings on price competition. Businesses get squeezed on price. How can they maintain profit margins? They do some squeezing of their own — hammering down suppliers and employees. We have an era where 'cost cutters' are corporate stars. Pah. And that's going to be true for quite a while, whether the economy is up or down. That's why I don't think we're on our way back to a time of company loyalty or job security. Instead, we will continue to live in the time of The Big Squeeze."

With that, I remembered how our conversation had started and I said, "Which takes us full-circle. We started out talking about fears and expectations."

My comment sparked a recollection in the old fellow. He stopped strolling and turned to me. "I haven't thought about this in years. There's a story that just came to me.

"It takes place back in the 1950s. Orson Welles was at the height of his powers as a motion picture director, and at the height of his arrogance as an artiste. This is not the cuddly Welles we got to know later, but the fiery, tyrant Welles. And the fiery tyrant decided that the CBS television network should turn one of his films into a television series. Welles arranged to meet with a CBS vice president, showed him the film, and then demanded his opinion.

"The vice president nervously replied, 'Great. I liked it. But do you think the average person would understand it?'

"The director looked down on him and said. 'Well, you understood it.'"

When Max told me that story we both laughed so hard that one of the slumbering businessmen woke up long enough to roll over and curse us.

Above and Beyond Average

Max then tied together what he'd been telling me on our walk: "We all want to be, expect to be, above average. But then there's all that overcapacity and all that fierce competition. You look around and see all the capable people moldering away in second-rate jobs, and eventually you conclude that it's hopeless. You realize that in today's world it would take a miracle to succeed, it would take magic."

It was time for another set of bruises from the old boy's grip on my shoulder. "That," he boomed away in the quiet, "is exactly what we need: magic. Isn't that what we want in our lives? We need to be amazed and astounded and surprised. And isn't that word one of life's grand compliments? Imagine having someone say about you,

'He took over and he worked magic.'"

His enthusiasm prompted more grumbling from those trying to sleep along the walls. We walked on before I asked, "How does this tie together with experimentation? Is the Hawthorne Effect the magic?"

The Secret of Achievement

This time he stopped and held one of my arms with both hands. "Hold on, young friend. There's a step we must not miss. I have to tell you about the moment that I realized the secret of achievement.

"One of the men I wrote about in my reporter days was Walt Disney — 'Uncle Walt,' he liked to be called. Now there was a man fizzing with intelligence. Someone once asked him his 'secret' and he said this: 'Do something so well that people will pay to see you do it again.'

"You can see how that credo fits his movies and amusement parks. But how did Uncle Walt do it? How do you amaze and delight people so much that they come back again and again?

"Those questions were on my mind when I took my kid to see the Disney movie *Snow White*, decades ago. There's a scene in that movie where Snow is standing beside a well. And she tells a flock of doves that it's a wishing well. She demonstrates, saying something like 'I wish my prince would come.' Then we see her from the bottom of the well, right through the water. We watch her face, shimmering in the surface of the water, as drops of water fall into the well and create ripples moving out. Now imagine drawing a shimmering face reflected in water that's rippling out in circles. Imagine how hard that would be, especially since this was long before computer animation.

"Why would Walt Disney have his artists devote all those hours of labor on that one little segment? If Disney had had a guy with an MBA on his staff, the guy would have said, 'Are you crazy? Cut that water nonsense. It doesn't add anything to the plot.'"

Then Max snorted out a laugh, having amused himself with a thought. "It just occurred to me that if the MBA and Walt Disney were talking today, the MBA would add, 'And while we're at it, why do we need seven dwarfs — why not six?'"

Before I could react, Max continued intensely, "But Uncle Walt was not an MBA; he was an artist. And the water scene stayed in the movie. Why? Because it hadn't been done. Because it was hard. Uncle Walt was showing off.

"And at that moment, sitting in the theater with my daughter, I realized what genius was all about. That scene

"And while we're at it, why do we need seven dwarfs -- why not six?"

in *Snow White* whispers the secret of achievement … "

And he dropped his voice so low that I had to bend in to hear him: "It's better than it has to be."

He resumed walking before he finally added, "True achievement is something better than it has to be. It's not just good, it's amazing. It's magic."

I understood what he was saying, but this piece of advice did not seem particularly helpful. I was hoping for something practical and he was off on "genius" and "magic." Who was I to come up with those two ingredients?

The Logic of Being Different

When I expressed my disappointment, he interrupted. "Don't be impatient," he said. Then he stopped himself. "No, forget that one: Be impatient. But remember what we said about the flute player, Jean-Pierre Rampal. He wanted to be 'better than perfect.' And now we look at Disney and see him being better than he has to be. Both those seem illogical, right? But they aren't, they are just part of a different logic. Maybe we should call it the Logic of Being Different."

Next he declared, "All of this experimentation may sound like hard work. Sometimes you'll be tempted to hide from it among the flock of the mediocre. That's when you need another bit of the Logic. Remember this: Mediocrity is hard. Creativity is easy."

He pointed once again at the people sleeping around us. "All of these folks want to be above average. And they are. They are top sales people and engineers and auditors. The problem is that there are so many above-average people that they are ordinary."

More of the Different, Less of the Same

"The tragedy is that because all of these people want to be better, they do the only things they've been taught about success — to try harder and think positively. So they press grimly forward, putting in more and more hours, doing more and more of the same.

"That's the standard approach: Do more. And it works. In fact, think about it: Isn't that the strategy of many of the Asian companies? If your product has ten defects per thousand, ours will have five. If your product has three gizmos, ours will have six. They are the masters of incrementalism. It worked. It worked in the auto industry until the number of defects approached zero and the Americans caught up."

We had by now walked back to where we had begun, and Max rummaged in his papers and came out with a cartoon by William Hamilton. It showed an executive talking on the telephone. This was the caption: "Look, Margo, I do not brag about how many hours I put in and I haven't got time to argue about it because, as you know, I put in a hundred hours a week here."

This time he didn't bark out a laugh. Instead he almost whispered. "That's the life being lived by most of the people in this airport. A dreadful, soul-deadening competition to see who can put in the most hours."

A New Game

Then he said, "Who wins that competition? No one. It's a game where everybody loses. So pick a new game."

"Does one really exist?" I asked.

"Sure. It's experimental . . . of course."

"Of course."

Chapter 11

By that time I felt I'd had enough philosophy. I told Max I was ready to turn practical and learn how I could implement this experimental magic.

He responded, "You've earned it, kiddo. We're almost there. The biggest challenge is to get your mind open. Once it's open, the ideas just pour into it. That's exactly what we want for you and your career."

Being Worthy of Great Ideas

That launched us into a discussion of how ideas come about. He reminded me of two of the stories he had already told me: Coca-Cola and Levi's jeans. "There were two of the greatest product ideas of all time, just handed over as a gift of the cosmos to John Pemberton and to Levi Strauss. Just handed over."

I made some crack about how lucky they were. Then the old boy surprised me once again with this: "If the cosmos were to give you a fabulous idea, would you be worthy?"

I was glad he didn't wait for my answer. "Pretend for a minute that you are John Pemberton, a pharmacist in Atlanta," he said. "And you go in the back room and there are two of your employees and they've gotten out a drug you're working on, and they are taking it for fun. What would you do? Honest. What would you do?"

I told the truth: I probably would have gotten angry and fired them on the spot.

Next, he asked me to put myself in the shoes of Levi Strauss. I'd just gotten off a long boat trip and was walking through San Francisco carrying all my belongings plus rolls of tent canvas, eager to go and search for gold. And a prospector comes up to me and says, "Got any pants for sale?" What would I do?

Again I had to be honest. I supposed I would have said irritably, "No, I don't have any pants for sale. Are you blind? I've got tent canvas."

And so I could see how I would have let two of the greatest products of all time slip right past me.

Mistakes — Gifts of the Universe

Next Max Elmore told me, "I'm sure you've had dozens of great ideas pass before you. It isn't easy being open-minded. In fact, our culture admires 'single-mindedness.'

We are so afraid of making mistakes that we don't bother to see that they are gifts of the universe."

Creativity Spilling Over

"A number of years ago I went to 3M headquarters," Max told me. "And I met with some executives there, including some who worked on the Post-it note. Everybody knows that story about the glue formula gone wrong, right?" I said that I did.

"So when I was there," he continued, "I asked one of the 3M people if they had noticed how often a new product was the result of some mistake. She said, 'Not only have we noticed, but when we put together a batch of product development stories we were embarrassed. It makes us look like a bunch of klutzes. So we've been trying to decide if we should even publish some of the stories.'

"I did what I could to convince her to go ahead. People need to know that ideas are always some kind of fluke. But she argued that because 3M spent so much money on research — something like a billion dollars a year — they ought not to look like they were just bumbling around.

"But that is what 3M gets for its billion dollars — a lot of open-minded people waiting to stumble across a great idea. The danger isn't embarrassment; it's missing wonderful opportunities."

Discovered and Recovered

Next, Max told me he'd once read about a man named Dr. Don Cooper. Max found Cooper's experience so intriguing that he'd tracked him down in Oklahoma to learn more.

Cooper has had a successful career as a physician. Besides his regular work, he's been on the President's Council on Physical Fitness and team doctor for the Oklahoma State football team. A successful guy. A guy who could have invented CPR — but didn't.

Here's how Max told the story: "It was the early 1950s and Don Cooper was an intern at a hospital in Kansas City. A man arrived at the hospital, complaining of various pains. When the young Cooper attempted to examine his new patient, the man turned angry and uncooperative and Cooper decided the patient really needed a psychiatrist. Checking with a senior member of the staff, Cooper was instructed to give the man an injection of a tranquilizer and then move him to the psychiatric ward.

"Cooper managed to get an IV started, but his patient was becoming more suspicious and less cooperative. The next medical step was to inject the tranquilizer through the IV. This needed to be done with care, for injecting it too quickly could be lethal. The young intern had just gotten the injection started when the patient decided to get up and leave. In the scuffling that followed, Dr. Don Cooper injected the entire dosage of tranquilizer into the man, who fell back and died!

"The panicky young intern listened to the man's chest: Nothing. No heartbeat. At that time there was no procedure for restarting a heart. In other words, the young Dr. Cooper had just killed one of his first patients.

"As the man lay dead on the table, Cooper's emotions swirled. The man was dead. Cooper's career possibly ruined. In desperation and in anger, Cooper struck the man, punched him in the chest, right above the heart that had stopped beating.

"The dead man coughed. Cooper, astounded, listened again to the man's chest. There was a heartbeat — irregular, but a heartbeat. So Cooper punched him again. And the man's heartbeat returned to normal."

I asked, "That was it? The patient was okay?"

"Well, when the man awoke he said, 'I feel better, but, boy, my chest hurts.' And Cooper said only, 'Maybe you've got pleurisy.'"

Max chuckled, then said, "Funny story. Sad story. It had a happy ending for the patient, but a sad one for medical science. When I asked Don Cooper if he had written up his experience for a medical journal, he said, 'Are you kidding? I killed the guy. I wanted to keep it quiet. It was thirty years later before I'd talk about it.' And it took nearly a decade for CPR to be rediscovered and adopted as a standard procedure."

Try Everything

Thinking about Coca-Cola and Levi's and CPR reminded me of our discussion of "flukology." I reminded Max of that expression and then said, "So what you're telling me is that I simply should sit back, wait, and watch and then a great idea will appear."

He grinned at me, holding up his hand and counting off his responses on his fingers. "The answer is no, no, yes, yes: no to sitting back, no to waiting, yes to watching and yes to great ideas appearing. And one more yes we shouldn't forget: yes to experimenting with the new idea.

"John Pemberton wasn't sitting back, he was experimenting with headache remedies and dozens of other new products. The fluke wasn't that he invented something but that he invented a soft drink instead of a drug. And Levi Strauss wasn't sitting back, he was traveling across the country and he was seeing what he could sell. Compare those two to Don Cooper, who watched a great idea literally come to life before him, but who was unwilling to acknowledge it and experiment with it."

I proclaimed, "So the conclusion is not, 'Do nothing and a great idea will come to you.' The conclusion is, 'Do everything because you never know where the great idea will come from.'" I must have gotten it right because I was rewarded with another set of bruises from his squeezing of my knee.

The Innovation Option

Next I said, "I hate to be a pest about this, but how could this possibly apply to me? I'm just a corporate employee."

He made a show of rolling his eyes and groaning. "That's what everybody says: that it's easy for someone else. Sure, they have the time; well, yes, they have the money; of course, they have the contacts." Then he barked in the dark corridor, "No." Then, nearly whispering, he added, "What they have is hindsight. Innovations seem simple, after the fact."

Max let me ponder that for a moment, then continued: "Let me give you an example that will make you see just what I mean. Let's take the case of a men's clothing store in San Francisco. Their business had declined to the point that they were losing money. You think you have troubles as a corporate employee, but imagine working sixty hours a week and losing money — that's like paying your company to let you work there.

"The owner was ready to give up. But he decided to talk to a man I met named David Wing, a small-business consultant. The shopkeeper didn't want to spend a lot of money, but he wanted a turnaround. Sounds impossible, right? Then this David Wing comes in and after spending some time in the store, tells the owner to:

1. Move everything around in the store.
2. Open at seven-thirty a.m. instead of ten a.m.
3. Buy a big fish tank.

"You'd think this guy was loony, wouldn't you? But here's his logic: There was a lot of pedestrian traffic passing in front of the store. The fish tank, just inside the store, would catch the eyes of passers-by. Why a fish tank?

"'Because,' Wing told me, 'I had never seen a fish tank in a men's clothing store.' In other words, it was different just for the sake of being different. And it inspired the store personnel to a lot of creativity. For instance, they had a mannequin up on a ladder bending over the tank so it looked like the dummy was feeding the fish.

"Next there was an idea of shifting around all the

merchandise. This was done simply to make the store look different. If you take all the same merchandise and put it in different places, a customer sees different merchandise.

"Finally, we come to Wing's idea of opening the store earlier. The plan was that businessmen on their way to work might just spot a tie they wanted for a big meeting that day, or realize they needed a heavier coat or an umbrella."

Thirty-Percent Improvement

"Their business went up thirty percent immediately and the store was once again making a nice profit.

"Looking back, the owner felt he had few options — he couldn't afford to slash prices or have a big ad campaign or even to remodel the store. Yet he discovered there were plenty of things he could still change, and all at once. The longer hours might have increased business five percent. Same for moving the merchandise around or adding the fish tank. But doing all three at once had a Hawthorne Effect on potential customers — who noticed something was happening and wanted to check it out. And I bet the staff felt the Hawthorne Effect, too; they were part of the experiment. That would mean that morale improved, and the quality of customer service would follow.

"And notice the coincidence that sales in the clothing store went up thirty percent, the same increase as in the Hawthorne plant experiments."

This prompted me to ask, "So it's always thirty percent? It's not twenty or forty percent?"

"It could be twenty or forty percent. It could be zero or two hundred. Still, that thirty percent number sticks with me. It's a solid improvement, a worthy goal. No, it isn't hitting the lottery. But why wait for a once-in-a-lifetime, Coke or Levi's kind of idea? If you do, you won't have any experience at spotting ideas and you'll have missed dozens of other innovations along the way."

A Reputation as an Experimenter

Nudging me with an elbow, Max Elmore said, smiling, "So, my young friend, you could do with a thirty-percent increase in your productivity and your income, right?"

I gave the expected answer.

"Well, change all that you can. Change enough so that people notice that you're changing. Arouse curiosity. Get a reputation for being an experimenter and people will bring their ideas to you."

I understood what he meant; still, I couldn't envision how I could change very much. My company had a lot of policies and guidelines. How much freedom did I have?

Not much.

The Idea Wellspring

In response, he told me, "I'll tell you how I generate ideas. Then, if you want to try it — and if they ever get this airport open — I have an assignment for you on your flight home. Make three lists.

"First, write down every mistake you've made in your career. Listen to what the universe has been whispering to you, and you may have some career CPR waiting to be discovered.

"Next go to problems. List every annoyance you have in your job, plus every complaint you've heard anyone else make. If you want to be a hero, solve someone's problem.

"Finally, make a list of everything you do at work. If you're going to change everything, start with a list of what everything is."

I wasn't sure what he meant by this last list. When I asked, Max responded by saying, "Right now, tell me one thing you do at work."

"Write reports summarizing sales data for top management."

"So 'Report Writing' would become one heading on the list. But add plenty of detail, make subheadings, like in an outline. Put on the list how you do your reports — all the steps along the way. When you write them; where you get the information; how you distribute it. You could have ten or twenty or fifty items under that one heading.

"Once you're done, you're not done. You're never done. Keep changing your lists. Keep all three handy and read them every day. The lists are the materials from which you can build a thousand new ideas. Remember: You couldn't invent Levi's jeans unless you were thinking about what to do with some leftover tent canvas."

Chapter 12

M ax sat, rubbing his hands together, being a ham. "Let's start with the easiest one, the List of Problems or Complaints."

Family Problem, Career Opportunity

Then he did what he always did: told me stories, and the one about the late Claude Olney was my favorite. I especially appreciated that story because many years before I had purchased Olney's course on improving school performance called *Where There's a Will, There's an A*.

If you aren't familiar with them, Olney's programs were a great marketing success story — they sold over a million copies. But did Olney sit down and decide to become wealthy selling a video course? No. He sat down and tried to solve a family problem.

"Olney's problem was," Max explained, "that his son got turned down for admission to Arizona State

University, the same school where Olney taught business law. Although he helped his son get a provisional admittance, the experience forced him to admit that his son needed help. So the professor took it upon himself to go to the library and check out a stack of books on studying and schoolwork. He was disappointed. The books gave the same dreary advice: plenty of study time and turning off the television and so on."

That part of the story certainly brought back memories for me. What student hasn't gotten the same lecture? "But," Max continued, "Olney knew that simply studying harder and longer was not going to work for his son. He decided instead to put the advice book aside and try to figure out what his own top students were doing to get high grades.

"He began to notice where the best students sat in class, how they studied, what memory tricks they used. And he began to ask other professors for advice and he began to read articles on education. He learned, for instance, the importance of the appearance of exam papers. A professor at Northwestern had once asked his wife to rewrite, word for word, some messy student exams. He gave the papers to his graduate assistants to grade. The papers written with his wife's gracious penmanship all received A's. The sloppy originals received mostly C's. A cosmetic change, sure. But one that would help his son.

"And here's another example: Olney observed that, on exams that required calculations, a lot of the mathematical errors were simply the result of getting the columns confused. Neat, straight columns of numbers decreased mistakes.

"Olney was able to give his son dozens of practical suggestions. And his son responded by making the dean's list and graduating with honors."

Max stopped here long enough to point out that the labors of Olney and his son resembled the Hawthorne experiments. They worked together, trying dozens of techniques, all at once, surprising each other with the results.

"But," he continued, "there was a parallel story unfolding, that of turning the study ideas into a business. Olney had been sharing his ideas with his students, who were, like his son, improving their grades. Next Olney turned his new knowledge into a seminar.

"As should be obvious, Claude Olney was an experimenter. Next he decided to try offering the course to high school students. He tried recruiting students by direct mail. He tried press releases. Most of his experiments failed in the sense that they didn't do what he hoped they would, but he made himself available for good fortune to find him.

"He wrote to a columnist at the *Chicago Tribune,* asking if he'd be interested in writing about his seminars. He was. The column generated over three thousand orders.

"Then it happened that Olney sat next to a former student at a wedding and mentioned, just chatting, what he'd been doing with his video seminars. The former student repeated the conversation to a friend whose daughter was struggling in school. This friend-of-a-friend bought one and it worked for his daughter. The happy dad happened to be in the 'infomercial' business. Soon the program was being offered on television and selling thousands of copies a week."

Then Max Elmore added a story within the story: "And where did that great title *Where There's a Will, There's an A* come from? Olney had been using the name *How to Get Better Grades in College.* It was his daughter, then a high school student, who suggested the new name."

Large Effort, Little Fluke

Max paused there. "So most people would meet Claude Olney and think, 'Now there's a lucky guy.' Right?"

"Of course. He was lucky," I insisted.

"Okay. But don't stop at lucky. He had a problem to

solve, and he put in a lot of effort to solve it. How many of us would commit that much effort to helping our children? And if we did, how many of us would go past the first good idea and come up with dozens more? I can't think of a success story that doesn't involve an extraordinary effort masquerading as a little fluke.

"So that leads us to the question: Is 'effort' the secret of success instead of 'luck'? Well, yes and no. There is never a single element of success. Look at all of the elements of Olney's story. Being a good father. Solving a problem. Rejecting the traditional approach and the conventional 'study harder' advice. Being an experimenter. Trying everything. And then, being open-minded, recognizing that he had something worthwhile. And next, becoming a marketer, starting the experiments all over.

"Here's a good question: Did Olney find the great ideas or did the ideas find him?"

The Flow of Ideas

"Well," I began, hesitantly, "he went and found some existing ideas, like the one about rewriting exam papers. And then some ideas found him, like the name of his videos and the 'infomercial' connection. So the answer is, both."

He started applauding. "Right, kiddo, right."

Next he said, "Olney put himself into the flow of ideas. You just have to start. There's no way to be sure which ideas will end up being good ones and which ones won't. You just start with as many as possible. And those ideas will attract other ideas. Once you get started, new ideas will come to you and jump on."

And then he was off to the next story.

Velcro

"Speaking of ideas coming to you," he said, "do you know how Velcro was invented?" I said, "sort of, something about burrs and pants." Max plunged ahead.

"A man named George de Mestral walked home through the woods. This is in Switzerland in the 1940s, by the way. When he got home he found cockleburs clinging to the cuffs of his wool pants. He knew it was useless to try to brush them away. As he plucked them off, he wondered how they did it, how they attached themselves so securely to his trousers.

"When he put the burrs under a microscope, he saw that each one was a ball of tiny hooks. His wool pants were made of loops of thread. The hooks of the burr latched onto the loops of thread in his pants.

"George de Mestral set about the work of re-creating

the cocklebur's system, using fabric. It took years of experiments, but eventually he was able to do what the cocklebur did. He gave his new product the brand name Velcro, taking the first syllables of two French words — velour for 'velvet,' and crochet for 'hook.'"

As was his habit, he asked me what we could learn from de Mestral and Velcro. I said that he had taken a problem and made it into a business. "Right," Max replied quickly. "And now for a bonus question: Did George de Mestral solve the problem?"

I replied, "He never did solve the problem of getting cockleburs off his cuffs, did he?"

"When I suggested that you make a list of problems," Max continued, "that wasn't meant to narrow your creativity but to expand it. People are always saying, 'A problem is an opportunity.' What they usually mean is that every problem is an opportunity to demonstrate how well you handle problems. You decide to 'solve' problems, to 'tackle' or 'overcome' them. But Mr. Velcro didn't 'solve' the cocklebur, he befriended it. It's like a saying they have in India: 'If you're going to live by the river, make friends with the crocodile.'"

I smiled at the image, but I was shaking my head, not certain what he meant. Which led to another story.

Dave Thomas and Wendy's

"One day I was visiting with Dave Thomas, the fellow who founded the Wendy's hamburger chain. We started talking about how he came up with new menu items. He told me that he had originally decided against offering salads at Wendy's, but now his company sells more salads than any other chain.

"Wendy's was buying lots of lettuce to put on their sandwiches, but they couldn't use the middle parts — the 'hearts of the lettuce' I think they're called — so they just threw them away. Thomas decided that he ought to be able to do something with all those leftover 'hearts.' He wondered, Was there someone who would buy them? What could somebody use them for? Only then it occurred to Thomas that he had a major ingredient for salads."

The moral of the story was clear to me and I jumped in: "Once Thomas started to think how he could convince someone else to use the leftovers, he saw how he could use them himself. He befriended the problem and then it wasn't a problem."

"Exactly."

"But that can't always be true, can it? Surely there are plenty of problems you can't turn around and make into advantages."

Dick Fosbury and the "Fosbury Flop"

"True," Max replied smiling, "some crocodiles make better friends than others. But you still need to get to know them all. Here's an example of a problem that might be called 'insurmountable.' What if you want to be a high jumper and you don't have a great vertical leap? Now that's a nasty croc!" That line was accompanied by a brisk elbow to my side.

"Back in the 1960s," he continued, "there was a guy named Dick Fosbury who invented what came to be known as the 'Fosbury Flop.' He was the first high jumper to go over the bar backward, as in upside down.

"I got to meet Fosbury once, and I asked how he decided to go over the bar backward.

"He explained that he had learned to do the scissors-kick jump in junior high, where you go over the bar in what is almost a standing position. But, in college, his coach moved him on to the 'roll' — where you go over the bar knee and shoulder first, facedown. But once his coach realized that Fosbury didn't have a great vertical leap, the coach soon lost interest. That was when Fosbury began to play around with his old scissors-kick style.

"Fosbury noticed that when he missed with his scissors–kick, it was because he knocked off the bar with his rump. This led him to experiment with lifting his seat when he jumped, and that meant leaning backward. Soon he was leaning so far back that he was no longer doing the scissors–kick, but had invented a style all his own."

Max looked at me with raised eyebrows, expecting me to offer a conclusion.

"That one seems easy enough," I said. "Fosbury wasn't a great leaper. That forced him to experiment with other styles. Which led him to the next problem — getting his butt out of the way. And then the experiments began again."

"Right once again. You could say that Fosbury's mediocre vertical leap was his enemy. Then again, it was his friend.

"And before we leave this story, there's one other thing I should pass along: Fosbury told me he didn't consciously plan any of his innovations. He simply paid attention to his problem — to what part of his body hit the bar — and then concentrated on getting it out of the way.

"That's important because I don't want to leave you with the impression that these problem-solving discoveries are always a triumph of analysis or cool calculation."

Leading Problems

Max then shook his head, his way of interrupting his own

train of thought. "Enough of famous people," he suddenly proclaimed. "I don't want you to think that I have to search history to find these examples. Let me tell you a story closer to your own situation, and see what we can conclude about problems becoming a career …

"Kathleen Duffy was director of research for an executive search firm in Phoenix. When the company moved to California and she chose not to go along, she ended up unemployed. So she began offering her recruiting skills to local corporations, billing at an hourly rate rather than a percentage of the new employee's salary. Working from home, over the phone, she'd find qualified employees, screen them, and persuade them to consider a job with her client company. She soon had more work than she could handle, and started adding researchers. Soon, Duffy Research had grown to a couple of dozen employees, all working from home except for one office manager.

"The company grew because Duffy stumbled into a niche — managers who didn't want to hire a headhunter, or couldn't afford one, but who didn't have the time or temperament to do their own cold-call recruiting. She fell into something close to the perfect job, one that makes her a salary into six figures.

"Not only was Duffy free to create her own job, she

has been able to share her wealth of freedom. Thanks to the success of the company, her husband was able to quit his job and open his own company as an illustrator. Duffy also hired her next-door neighbor — at the time, an executive assistant — who eventually got to the point of earning a six-figure salary, enabling her husband to quit his job to go in the collectible toy business."

Max gave me his look that meant, What can we learn from this? I responded by saying, "She never set out to invent a business; she never even consciously worked at analyzing her problems. She simply let her problems — and the marketplace — lead her."

Max nodded, acting impressed, and added, "There's a lot to be said for following. In fact, I never thought of it this way, but that story reveals a Zen-like truth about many of the best careers: Search for work among the jobs that don't exist."

I wasn't sure about that formula at the time, but I've come to see that competition for great jobs either makes them exceptionally difficult to obtain, or else drives down the rewards. Creating a job where none exists can defy such supply-and-demand pressures.

However, sitting there in O'Hare I wasn't thinking about finding a job that didn't exist, I was still thinking about how to create a lucky problem: "But if someone — I mean 'me,' of course — if I don't want to wait around for problems to pop up, I could ask myself, 'What could go wrong?' and then, 'What could we do if those things did go wrong?'"

Max worked his way to his feet, just so he could do an elaborate bow. "The student teaches the teacher," he said, bent over from the waist.

Obvious Courage

Once I got Max to sit back down, he began reflecting on problem solving. "Remember this: The solutions look obvious in retrospect. Occasionally, the solutions are easy to implement, like adding salads to the menu at Wendy's. But more often than not, the solutions require plenty of courage. Claude Olney got his son into college even when he wasn't qualified — I'm sure there are people who told him that he was making a mistake, setting up his son for failure. And imagine how people must have scoffed at George de Mestral for spending years trying to re-create a cocklebur. And imagine how they laughed at Dick Fosbury and his 'flop.'

"Remember: It's easy to experiment, but hard to change. None of these people started with a grand vision and then worked toward it. These people were not goal setters or planners; they were adventurers."

Experimental Journey

Then Max concluded that part of the conversation with words that I will never forget:

"Each quandary is a call for experimentation. Each experiment a question put to the world. Each answer a journey. Let life plan the itinerary. Your job is to pack light and bring a camera."

That said, he told me he needed to "rest his eyes." I didn't dare object because by then it was after three in the morning. While he dozed, I frantically tried to write down all I could remember of what he'd said in the hours before.

Chapter 13

A while later — a lightening of the gray clouds told me a feeble dawn was underway outside — one of those motorized airport carts came by, making whooping noises, and my companion woke up. I suggested that we see if they had restocked any of the restaurants. As we searched I returned to his idea of using the three lists as guides to experimentation. We'd already discussed the problem list; I urged him to describe the next one, the Work List.

Creativity Caffeine

Max smiled weakly; he seemed tired. I told him that we could talk about it later, when he felt better. He said, "Talking is what will make me feel better. Creativity is pure caffeine. Let's come up with some ideas, just for fun." With that he stopped a security guard who was trudging past. He said to this guy, "I need to show my young friend how to work some magic. I want to show him how to pull ideas out of a hat."

The Idea Generator

The security guard must have thought Max was batty, but with nothing more to do but walk the halls, he shrugged and nodded an okay.

The old eccentric demanded of him, "Tell me this, where could you use some creativity in your life? What would you like to do better?"

The guard had no answer. Finally, after considerable prodding, he replied, "Okay, I got one. I could use some new excuses for being late to work. I've used up all my old ones."

I groaned inwardly, but Max seemed delighted. He told me to get out a piece of paper and a pen. He asked the guard to name the best excuse he had used lately.

"My hot water heater exploded," the guard said quickly. "One of the pipes popped loose. It took all day to get someone to come out, and then part of the next day too. I didn't even have to pretend to be sick."

My instinct was to resume our walk, but Max instructed me to start a list of pairs of words, with the first pair being *hot water heater* and *exploded*. Then he asked both of us to give him some excuses we had heard co-workers use — the first to pop into our heads, good or bad.

He kept us going till we had listed eleven excuses, ranging from reasonable ones like the hot water heater episode and one about having to give a police statement when a drunk driver was arrested, to awful ones like the woman who was two hours late for work because her dress got caught in her zipper, or the man whose dog got loose and headed down a busy street.

The result was a list of pairs of phrases, to which Max added numbers down each side, from 02 to 12. He showed us the finished list and said, "Not bad, but no Tabasco, either. The action starts when you add a pair of dice.

"This is how it works. Make two throws of the pair of dice. Say you get eleven and nine. These become the character and event for your new excuse."

It turned out that number eleven was the first half of "elderly aunt called" and number nine was the second half of "dog wandered off into traffic."

He continued: "So you take the 'aunt' and the 'wandering into traffic' and then you have the basis of a story. You might say, 'My elderly aunt's neighbor called. My aunt had taken off walking, right across the middle of the street, ignoring the cars. The neighbor brought her back home, but I had to rush over there, then wait till my mother could get over there to replace me.'"

We didn't have a pair of dice, so he asked us to pick pairs of numbers at random. Some gave us strange images — the pair *spouse* and *stuck in tree*, for example. Others were gruesome — like *dog* and *exploding*.

Max summed up the session by giving the list to the guard and saying to him, "You'll know that you've mastered the technique for producing ideas when you're turning out stories that have your boss saying, 'It must be true — no one could make up an excuse like that.'"

Harvesting Ideas

Once we had resumed walking, Max looked at me and inquired, "What did you learn from that?"

"That you can come up with new ideas in a hurry if you know how to do it."

"Good. And what was our source material for new ideas?"

"Old ideas."

"Exactly. New ideas are old ideas in new places. Some of those excuses were terrible, weren't they? But it didn't matter — the smelliest garbage makes the best fertilizer." He then laughed at himself. "I don't know if that's true or not, but it sounds wise, doesn't it?"

He continued: "So we're going to go around and pick ideas, just like walking in the woods and picking wild berries. Are you ready for more?"

He began by recalling an earlier discussion, saying, "You remember that when we talked about making a list of what you do, I suggested that you make it very detailed?" I assured him that I did.

"Most people," he explained, "define their jobs too narrowly. A guy assumes that if he has the best engineering skills in town, he's the best engineer. But being a great engineer takes a lot more than the mechanics of the job. It takes the ability to sell ideas, to work with people, to lead meetings, to avoid idiotic meetings — there are a hundred skills involved.

"That's why keeping a list of all your duties is important. And you should keep redefining your job, expanding your list."

Muhammad Ali

"Okay. First example. Muhammad Ali. He was a good boxer, but there have always been plenty of those. How did he get to be a world hero? It began the day he realized that his job wasn't just to throw punches, but to draw a crowd. I'll never forget one interview with Larry King when the champ described the origins of his media personality.

"Ali had one of his early pro fights in Louisville. A

few days before, he appeared on a local sports talk show. On the same show was Gorgeous George, who was in town for a wrestling match at the same arena the night following Ali's fight.

"Ali politely gave standard diplomatic answers to the questions he was asked — the usual, 'I'm going to do my best' responses. Next the host asked Gorgeous George about his match. The wrestler began shouting, 'I'm gonna kill them! I'm gonna bring venom and menace and horror to Louisville Saturday night!'

"Ali fought his fight and won. But he noticed that although he drew four thousand fans, Gorgeous George drew thirteen thousand.

"That's when Ali started his outrageous behavior. Soon he was predicting the round of his knockouts, taunting his opponents, and bragging about how pretty he was. In other words, he brought the showmanship of wrestling to boxing."

Max summed up the Ali story by saying that the boxer "was smart enough to define his job broadly and smart enough to realize that he could learn from someone in another sport."

Next, Max turned his attention to the next step, transforming a list of old duties into a list of new experiments. "The reason we did the exercise with the security guard was to show you the pleasure of recombining ideas. I used to employ creativity tricks like that one to sit down and generate ideas. You might want to try it. On the other hand, what I like to do is take the list of what you do on one side, and the entire world on the other side."

60 Minutes

He moved on to an example that fascinated me. I've always liked to watch the television show *60 Minutes*. Max told me how the show came into existence.

"The show's producer, Don Hewitt, was trying to come up with a new program. He noticed that television had the nightly news, which was the electronic equivalent of a newspaper, and that it had documentaries, which were like nonfiction books. But it had nothing analogous to magazines. Hewitt's idea was to create a 'magazine' for TV. You can see that once you have that picture in your head — a 'magazine' on television — all he had to do was spend some time at a magazine rack to collect ideas about how to handle the program."

I responded, "That seems just like our excuse exercise — a simple recombination. Magazines and television."

"Exactly. And it happens all the time. We have 'tabloid television' — tabloid newspapers combined with television. And also notice that there are television-style newspapers — *USA Today*, particularly. All you have to do is take whatever you decide to improve and look around for ideas to borrow."

Twitter

"Oh, oh," Max added, bouncing, "here's another example, one a young guy like you can appreciate. Do you know who Jack Dorsey is?"

That sounded familiar, but I wasn't sure. "Was he the guy who helped create Twitter?"

"Righto. And if you're like me, when you first heard about Twitter you thought, 'What? Who would use that?' So that means it was a giant leap of imagination, right?"

Before I could agree, Max gave a hard, contrary "Ha!" and then explained.

"Turns out that as a boy, Jack Dorsey had a speech impediment. That helped make him something of a loner, one who spent a lot of time with computers, teaching himself to program them. And he was fascinated with trains and how they came in and out of the train yards near his home in St. Louis. From that, he began listening to dispatchers, using a police scanner. He liked how they communicated in short bursts, letting everyone on the network know where they were and what they were doing."

Max summarized the lesson this way: "Having heard the background on Dorsey, you look at Twitter and go, 'Of course. It's like dispatchers communicating. That's a small step.' So what at first seemed like a gigantic creative leap turns out to be a logical little relocation of an idea."

Starting the Innovation Spiral

Then, Max Elmore, advisor to presidents and CEOs, turned his attention to my little job. "You said before that one of your jobs is writing reports. As soon as you say that, you have something to work with. What if you didn't just write them? What if you recorded highlights on a podcast? And what about the data you report: What other measures could you include? What are other companies doing with their reports? How about Japanese companies? German companies? Spend your lunch hour online looking at how book authors and magazine writers report their data. Do that and you'll be lunching with Lady Luck.

"Don't forget what we said about ideas attracting other ideas. Get a few suggestions together, make up some samples, and take them to the people who get your reports. Ask them if they like any of the ideas. They'll be delighted that you asked. And they'll give you new suggestions.

"If you went to them and said, 'What could we do differently that would improve these reports for you?' chances are they couldn't give you an answer. But if you start the process, they have something to respond to. The same people who would have said, 'They're fine just the way they are,' will suddenly have lots to add. Most people don't have ideas; they have suggestions."

Chapter 14

That left us with just the last of the three lists, the List of Mistakes. Max smiled at the idea of that final list. "That one you must save for last. It is reserved for the advanced student. After you have worked with the other lists, come back to it. And when you return to your mistakes, you must put aside your emotions, your blame, your anger.

"There is a story that Joseph Campbell, the late writer and professor, used to tell, a tale of a samurai warrior sent to avenge the murder of his overlord.

"The warrior tracked down the murderer, then raised his samurai sword above his head to deliver the killing blow. The murderer, unrepentant, spit in the warrior's face. At that, the samurai turned away, sheathed his sword as he walked off, planning to return another day."

All of which made no sense to me.

"He walked away," Max explained, "because he had become angry, and he would have struck the killing blow out of his own rage. Instead, he chose to wait and return, so the blow would be of an impersonal act of retribution. And that is how you must look upon your mistakes, without emotion."

He'd gone over my head with this story, just when I thought I had the idea-generation process figured out. My confusion showed, because he backed up and re-explained.

"Your inclination will be to cover up your mistakes. You have to have the discipline to hold them up and examine them. You can't be embarrassed or resentful. That's why you have to come back later, like the samurai.

"Eventually it will happen that you do something silly … oh, maybe you'll put a password on a computer document and then forget what it was. Later, when you're finished being mad at yourself or the software company, you might just develop a piece of software that retrieves passwords. I met a fellow who was one of the first to develop and market just such a program. He stumbled across something useful. That expression says it all: stumbled across something useful.

"Get ready, my friend: What I'm about to say may take you a while to get used to. It might take a month or a year or a decade.

"We agreed before that when you befriend your

problems, they are no longer problems. Go far enough into a problem and you come out on the other side, the anti-problem. Well, looked at dispassionately, a mistake is just another problem. Same with failure: just a big problem. Go far enough into failure and you come out on the other side, the anti-failure."

He cupped a hand to his ear. "Did you hear that? It was the sound of one hand clapping." And once again he laughed his large laugh. "You'll make a good samurai one day, I promise you."

Eager for Morning

At that moment the security guard, the one with the list of excuses we had done together, came to tell us that the airport was about to reopen and if we went now, we could beat the lines at the ticket counters.

Before we went to our separate gates, old Max gave me this parting advice: "I wish for you that you find joy in your experiments. Be wealthy in ideas. Try everything. Be different tomorrow from the person you are today. Be one of the happy warriors, eager for morning."

*"Be one of the happy warriors,
eager for morning."*

Epilogue

That farewell was spoken nearly five years ago. I have talked several times since with Max, although I haven't seen him. He's been living in Australia, helping his son with a new company. I like to picture him out there, across the Pacific, experimenting.

Max Elmore should have written this book himself, as I reminded him every time we spoke. His response? "I've never spent much time looking in rearview mirrors." Still, I persisted. Finally he said, "You write it." So I penned this volume as a tribute to the man I met that snowy May night.

I don't try to hold myself up as an example of fabulous success, but for those who might be interested, I will tell you what happened to my career after that night.

On the airplane home I did what he had suggested — I made my lists. He was right: Once I wrote down everything that my job consisted of and kept a running list of every problem I encountered or heard of my co-workers encountering, it was a simple matter to come up with ideas for experiments. As soon as I started looking to be different, I saw ideas everywhere. For instance, I read Sam Walton's autobiography and got dozens of ideas from that book alone. And a book called *Walt Disney: An American Original* brought me dozens more. Now every novel, every television program, even every commercial can be a source of innovation. I'll say to myself, "Look at the recombination they came up with. Not bad."

Although I have never reached the level of being different every day, I am part of continual experimentation. In fact, as soon as I started experimenting, ideas came looking for me.

Within a matter of weeks after that night in Chicago, I had established a reputation as a doer, someone who made things happen, someone worth knowing. And I deserved that reputation, because ideas were flowing through me — I was truly alive for the first time in my adult life. Here's how it happened.

I returned to my not-so-special job as a corporate staff person. But I returned with a list of ideas, which I presented informally to my boss over lunch. I never said I wanted to change the department, just that I thought some experiments might be useful.

He took a few of my ideas and carried them to his

boss, who added a couple ideas of his own. Suddenly my boss saw me as a person who could help him impress his superiors. Within a couple of months I was promoted to a new department created just for me; I became manager of "Special Projects." In other words, I was given the job of being a full-time experimenter.

While my former peers were cranking out the same old reports, I was meeting with the company's top executives, talking about how we could make the company better. Soon I was traveling around the country in search of ideas, learning at an incredible rate.

Eventually I opened my own company, doing consulting. I knew that there were more consultants in the world than mosquitoes and that I would have to be able to solve some problems, to be different from the others. And so I came up with seven specialties I would offer. Guess what? Six of the seven flopped. But the seventh caught on, and soon I was getting projects from dozens of companies.

I'm proud to say that we never stopped experimenting. And because our clients knew that we loved to experiment, they came to us when they wanted to try something new. And when those innovative projects were finished, I'd suggest to the client that I write up an article on what we had done, making him or her a co-author. So we had a stream of notices in the trade press — free advertising.

And within our company, because all the employees knew I wanted their suggestions, they were eager to share ideas. And no one could get too down about any working condition because everything was permanently temporary. (We often reminded each other that "perfection is the first stage of obsolescence.")

I even came to understand the value of "making friends with the crocodile." Eventually the ideas of the "anti-problem" and "anti-mistake" made sense to me too. Now a business disaster is a chance to test ourselves, like the time I discovered I had botched a schedule on a project for a big new client. When I agreed to a deadline, I misread my calendar, got off by a month, and promised the finished report a month early — in less than eight weeks, instead of the normal twelve. Possible? Plain old hustle could make up a week or two, but that was all. That's when we realized that by combining two steps into one, we could save two or three weeks. So we not only made our deadline, we'd learned how to save time on all future projects.

No, I can't tell you that I had a fabulous idea, something comparable to the personal computer or the microwave oven. Instead, our company put to use hundreds of little ideas and we made ourselves leaders in our industry.

Looking back, I went from depressed to energized overnight. Literally overnight — that single night at O'Hare. And I've stayed that way. Now I fancy myself as having become a happy warrior, a warrior for change. Maybe you'd like to join me. Today. Right here. Right now. Just remember this: Change is hard; experimentation is easy.

Addendum: Instant Brains

Before I put Max's ideas into this volume, I had come up with a way to make use of my notes from that night in O'Hare. I summarized for myself some of the stories Max told me, and some others he prompted me to learn, and then tried to add questions to them to stimulate my thinking. This list I have come to think of as "Instant Brains."

Although the best ideas are often unpredictable flukes or coincidences, there are some occasions when you are called upon to generate new ideas quickly. Say a client says to you, "We need some fresh ideas for the new project — let me know if you can think of anything." And the next morning you deliver a list of fifty suggestions.

It's easy, if you remember that new ideas are old ideas in new places.

If you want to try it, start with a problem you need to solve, and then go through the list quickly, looking for immediate connections, making notes. Once you're through, you can then go back, looking for the next generation of ideas.

1. Coca-Cola was "discovered" when pharmacist John Pemberton found two shop boys drinking one of his headache syrups.

How can we engage the energy/creativity of youth?

Who should we let "play around" with our products/ services in order to reinvent them?   (Steve Wozniak developed the first Apple computer to show off for his fellow members of a computer club — is there a club or is there a kids/teens/school angle?)

Could we give our products to high school/college students to use and ask for suggestions?

What about the opposite — is there a senior angle? Should we be going to senior centers?

2. Sam Walton loved to experiment with promotions at his Walmart stores, especially placing products outside the store — for instance, he put a popcorn machine outside, then an ice cream machine.

How could we take our work outside?

What could we do so that passersby would see something different going on?

How could we manage our first impressions?

Sam Walton also turned his security guard into a "greeter" — is there an analogy here? How do we greet our users? (The owner of the St. Paul Saints minor league baseball team greets visitors as they enter, and so does a bubble machine.)

3. Cannondale bicycles are called Cannondales because the first day of the company's existence, an employee was sent to a pay phone to call the phone company to get a phone line. The clerk for the phone company asked for a company name. The new little business didn't yet have one. The clerk insisted that the account had to have a name. So the employee looked up and saw a sign proclaiming that he was at the Cannondale train station, so that's what he gave. Everyone liked it and it stuck.

Look around the room — what ideas are sitting there?

How did the products/objects you see around you get there?

Why did you choose the place you're in?

4. One of the most popular flavors of the China Mist Tea Company is "mixed berry." It was created when three leftover partial batches of fruit flavored tea were mixed together rather than being thrown away.

What are the leftovers from our endeavors that might

be used here?

What has been thrown away that could be salvaged for another use?

Dave Thomas of Wendy's was trying to find a use for the "hearts" of lettuce that the company was throwing away after using the rest of the heads for hamburgers. He asked himself, "Who could use these? Livestock feed, maybe? Someone who sells salads?" And that's when he realized that he could use it himself in salads, something Wendy's did not offer at the time. Wendy's now sells more salads than any other fast food chain.

The leftovers need not be physical products, of course, they could be unused or underutilized brains or hands or land.

Sam Schoen, the founder of U-Haul, built the company on the principle of "unused land and labor." He observed that gas stations had empty asphalt and employees who could take time from selling gas to rent trailers.

As a market researcher, I discovered a file cabinet full of old ad test results for the company's commercials. I put these in a database, so we could compare new test results to the old ones, and suddenly the worthless old data were valuable wisdom.

5. Most of our energies go to shoring up weaknesses, but

there is great leverage in pushing strengths. Jean-Pierre Rampal said, "Some nights I go out and play a piece perfectly; and then the next night, I go out and play it better."

What aspects of the endeavor are perfect? How could they be changed?

6. David Wing, a consultant to retailers, was working with a failing men's store in downtown San Francisco. He spent some time at the store and made three suggestions: (1) add a large fish tank, (2) move all the merchandise around within the store, and (3) open early enough to be available to men on their way to work. Part of the logic was to be different for its own sake — to make the store look different and so customers would see the merchandise differently (new context, new angles).

What can you do to move the project out of its normal context, to make people stop and look. This leads to the question,

Where do you NOT expect to find this current undertaking?

Plus,

What pieces can you move to a new place, or put in a different order?

What could be different just for the sake of being

different?

7. Claude Olney was a college professor who evolved his successful video program *Where There's a Will, There's an A* out of his attempts to help his son get better grades in college.

Who can you help? Who ELSE?

How does home or family life fit in?

How can work and home/family fit together to create a synergy?

Olney's program was successful because he tossed out the usual "study harder" approach and instead noted the habits of his own best students.

Who DOESN'T need help and why?   Who can we observe?

8. George de Mestral invented Velcro after observing how tenaciously cockleburs stuck to his wool pants.

Create a list of annoyances — your own and those who you want to help.

Befriend your problems — remember how Dave Thomas started selling salads: Whom could you sell your problems to?

9. Dick Fosbury won an Olympic gold medal with his

"Fosbury Flop," where he passed over the bar upside down. He stumbled upon this technique only after his coach gave up on him, leaving the young high-jumper free to revert to the old-fashioned "scissors-kick," which evolved into the "Flop" as he tried to get his butt out of the way.

What old-fashioned technique could be brought back?

What old designs?

How were related endeavors handled in "the old days"?

(Look for ideas in old magazines or newspapers. For instance, the head of a major Shakespeare festival told me that they keep a collection of fashion books from various eras and look through them to try to find new settings for classic plays.)

10. Muhammad Ali learned his outrageously boastful posturing by watching the raving of the professional wrestler, Gorgeous George.

What dramatic boasts could you make? (Thomas Edison set up press conferences to announce breakthroughs that hadn't happened yet — he wanted the pressure to perform.)

What outrageous predictions?

Who's getting the publicity in related fields and why?

Ali also realized that his job wasn't just boxing, but drawing and entertaining a crowd.

Redefine the endeavor — narrow it, broaden it.

How can you add value to your project by taking over some of the tasks of those you work with?

11. Don Hewitt created the television program *60 Minutes* when he realized that television had the equivalent of books and newspapers but did not have programming comparable to magazines. In other words, he combined two old ideas — television and magazines — to create a new combination.

Find other places/industries to look for ideas to mate with yours.

Are there ideas to be borrowed from the media?

How about from a nearby mall?

Each section of a drug or grocery store?

Now that you have created a list of possibilities, pick the one you like best and then try to pair it up with every other idea on your list — after all, a great idea is usually just two good ideas put together.

Acknowledgments, Sources, and Suggested Reading

One advantage of my years of working in market research is that I learned humility. Many times when I was given a new product or new advertising campaign to assess, I'd think to myself, "Whoa! They're actually going to pay money to test this loser?" And as often as not, it would become the hit of the year. The same was true for my predictions about market winners. So it was natural for me to accept the wisdom of Fats Waller: "One never knows, do one?"

It's in that spirit that I write weekly newspaper columns. Each week is a little adventure for me, and I hope for those who read my column. Along the way I have never missed a chance to meet anyone who was doing something different, something interesting. Many of the tales of those different and interesting people have been given a place in this book and I thank them for their time and their wisdom.

In addition to the innovation stories told to me directly, for nearly two decades I have been collecting ones already put to paper. I turned to that collection for examples to include here. As I assembled my list of sources I saw that these same books are ones I would gladly recommend to anyone looking for additional reading:

A major influence on this work is the classic review of the Hawthorne experiments, *Management and the Worker*, by F. J. Roethlisberger, William J. Dickson, and Harold A. Wright. The original copyright is for 1939, but the copy I was able to get my hands on was printed in 1961 by Harvard University Press.

Three other authors and/or books mentioned in the text were Burton Malkiel and his *A Random Walk Down Wall Street* (Norton, 1990); *Breakthroughs! How the Vision and Drive of Innovators in Sixteen Companies Created Commercial Breakthroughs that Swept the World* by P. Ranganath Nayak and John M. Ketteringham (Rawson Associates, 1986); and *Judgment Calls* by John C. Mowen (Simon & Schuster, 1993).

The last of these not only helped refine my thinking about what I call the "achievement lottery," but when I interviewed Dr. Mowen, he graciously arranged for me to talk with Dr. Don Cooper, the fellow who punched his dead patient in the chest.

Books where I first encountered stories of innovation were:

For Apple computers: *Accidental Empires* by Robert X. Cringely, (Addison Wesley, 1992); for *60 Minutes*: *Close Encounters* by Mike Wallace and Gary Paul Gates (William Morrow, 1984); for Orson Welles: *My Saber Is Bent* by Jack Parr (Simon & Schuster, 1961); for samurai: *The Power of Myth* by Joseph Campbell (Doubleday, 1988); and for Muhammad Ali: *Tell It to the King* by Larry King (Putnam, 1988).

The development stories of familiar products like Coca-Cola appear many places, often in corporate literature. Two collections of product stories that I have found helpful are *How the Cadillac Got Its Fins* by Jack Mingo (Harper, 1994); and a charming book for kids called *The Invention Book* by Steven Caney (Workman, 1985).

As for the creation of the book you are holding, there are many people to thank. First I owe a debt to those who helped launch my newspaper column and in doing so kept me writing. My thanks to the man who had the courage to run my work first, Jim Fickess. And many thanks to the big-city editors who adopted my column before it was syndicated, especially Phil Gaitens of the *St. Louis Post-Dispatch*; Larry Werner and Randy Salas of the

Minneapolis Star-Tribune; and John Genzale and Richard Heimlich, who made possible the column's syndication by King Features. Thanks also to the thousands of readers who have taken the time to write or call me. It was that response that made me decide to write this book.

Bob Nelson, author of the delightful and practical book *1,001 Ways to Reward Employees* (Workman, 1994), introduced me to the Margret McBride Literary Agency. Margret, and her assistants, Kimberly McBride and David Fugate, made this a far better book than it otherwise would have been, and then got the manuscript to the book's first editor, Henry Ferris, whose editorial wisdom made it better still. Many thanks to Steve Chandler for his generous Foreword to this edition and for introducing me to the insightful and delightful Maurice Bassett. It was Maurice who had the courage to take a chance on re-releasing this book and who brought in the charming and wise Michael Pastore as editor. They brought an unexpected exuberance and collegiality to the publishing process.

Finally, my thanks to those who talked through ideas with me and/or who read and improved the early drafts of the manuscript: Joel Dauten, Jim Fickess, Mark Nelson, Ted Dunham, Trevor Dauten, Lloyd Murphy, Mary Baroni, Tom Hendrick, Sara Harrell, Ray Karesky, Dan

Schweiker, Edvard Richards, Jeanne Winograd, Len Harris, Mark Quale, Steve Patchen, Fred Brownfeld, Sandy Dauten, and especially, to the late Roger Axford, whose charm and laugh inspired the personality of Max.

About the Author

Dale Dauten has been researching achievers and innovation since his days at Arizona State as a National Science Foundation Fellow, and at Stanford University's Graduate School of Business.

As founder of The Innovators' Lab®, Dale has consulted with many leading firms, including Caterpillar, Georgia-Pacific, Penske Automotive Group, General Dynamics, and NASA.

Dale's books have been published in a dozen languages and have developed a large following worldwide, especially in Japan. He also co-writes newspaper columns with J.T. O'Donnell, nationally syndicated by King Features.

Dale and his wife, Sandy Dauten, live in Tempe, Arizona.

MAURICE BASSETT

Publisher's Catalogue

The Mahatma Gandhi Library

#1 Towards Non-Violent Politics

* * *

The Prosperous Series

#1 The Prosperous Coach: Increase Income and Impact for You and Your Clients (Steve Chandler and Rich Litvin)

#2 The Prosperous Hip Hop Producer: My Beat-Making Journey from My Grandma's Patio to a Six-Figure Business (Curtiss King)

#3 The Prosperous Hotelier (David Lund)

* * *

Devon Bandison

Fatherhood Is Leadership: Your Playbook for Success, Self-Leadership, and a Richer Life

Is it ever too late for a miracle?

If Joe Dolan has his way he won't be around for Christmas. While the streets are bustling with holiday shoppers and good cheer, Joe sits alone in a bar and contemplates the end. Estranged from his daughter, divorced from his wife and cheated out of his business, Joe drowns his memories in whiskey and regret, convinced this will be his last night on Earth.

But . . . what's this?

A mysterious note scrawled on his cocktail napkin hints that someone knows what Joe has in mind—and wants to offer him a second chance. And Joe, on the brink of ending it all, decides to take it.

Available from Amazon, Barnes & Noble and other bookstores.

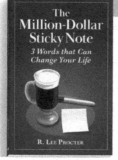

The
**Million-Dollar
Sticky Note**
*3 Words that Can
Change Your Life*

R. LEE PROCTER

"Heartfelt, hilarious and shockingly deep."

~ Harris Orkin, award-winning
author of *You Only Live Once*

"A bull's-eye to the heart."

~ Gill Reavill, author of
This Land Is No Stranger

Michael Bassoff

RelationShift: Revolutionary Fundraising (Revised Edition)
(Steve Chandler and Michael Bassoff)

Roy G. Biv

*1921: A Celebration of Toned 1921 Peace Dollars as
Numismatic Art*
*Dancing on Antique Toning: A Further Celebration of
Numismatic Art*
Dancing on Rainbows: A Celebration of Numismatic Art
Early Jack: The "Lost" Photos of John F. Kennedy
Early Jackie: The "Lost" Photos of Jackie Bouvier

Sir Fairfax L. Cartwright

The Mystic Rose from the Garden of the King

Steve Chandler

*37 Ways to BOOST Your Coaching Practice: PLUS: the 17
Lies That Hold Coaches Back and the Truth That Sets Them
Free*
50 Ways to Create Great Relationships
Crazy Good: A Book of CHOICES
CREATOR
Death Wish: The Path through Addiction to a Glorious Life
Fearless: Creating the Courage to Change the Things You Can
How to Get Clients: New Pathways to Coaching Prosperity
*The Prosperous Coach: Increase Income and Impact for You
and Your Clients (The Prosperous Series #1)* (Steve
Chandler and Rich Litvin)

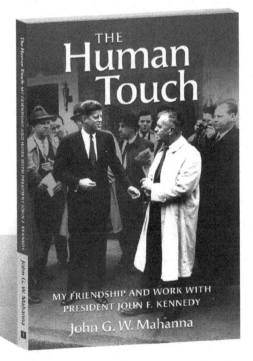

M. K. Gandhi

Towards Non-Violent Politics (The Mahatma Gandhi Library #1)

James F. Gesualdi

Excellence Beyond Compliance: Enhancing Animal Welfare through the Constructive Use of the Animal Welfare Act

Janice Goldman

Let's Talk About Money: The Girlfriends' Guide to Protecting Her ASSets

Sylvia Hall

This Is Real Life: Love Notes to Wake You Up

Christy Harden

Guided by Your Own Stars: Connect with the Inner Voice and Discover Your Dreams

I ♥ Raw: A How-To Guide for Reconnecting to Yourself and the Earth through Plant-Based Living

Curtiss King

The Prosperous Hip Hop Producer: My Beat-Making Journey from My Grandma's Patio to a Six-Figure Business (The Prosperous Series #2)

David Lindsay

A Blade for Sale: The Adventures of Monsieur de Mailly

Rich Litvin

The Prosperous Coach: Increase Income and Impact for You and Your Clients (The Prosperous Series #1) (Steve Chandler and Rich Litvin)

David Lund

The Prosperous Hotelier (The Prosperous Series #3)

John G. W. Mahanna

The Human Touch: My Friendship and Work with President John F. Kennedy

Abraham H. Maslow

Abraham H. Maslow: A Comprehensive Bibliography

The Aims of Education (audio)

The B-language Workshop (audio)

Being Abraham Maslow (DVD)

The Eupsychian Ethic (audio)

The Farther Reaches of Human Nature (audio)

Maslow and Self-Actualization (DVD)

Maslow on Management (audiobook)

Personality and Growth: A Humanistic Psychologist in the Classroom

Psychology and Religious Awareness (audio)

The Psychology of Science: A Reconnaissance

Self-Actualization (audio)

Weekend with Maslow (audio)

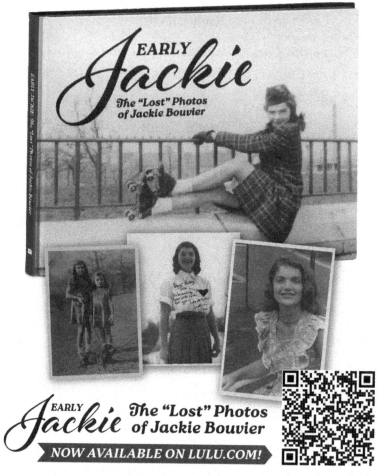

A full-color, coffee-table style photo album of 67 previously unpublished and seldom-seen photos of Jackie Kennedy from her childhood and teen years.

There must be hundreds of thousands of photos of Jackie Kennedy (1929-1994), our much-loved First Lady, either with or without President John F. Kennedy, but what you are about to experience in *Early Jackie* is strikingly different from the well-known and classic Jackie photos. These are the "lost" photos of Jackie from when she was known—prior to marriage—by the name of "Jackie Bouvier."

Jackie Kennedy lovers everywhere will delight in owning this remarkable, full-color photo album!

R. Lee Procter

The Million-Dollar Sticky Note: 3 Words that Can Change Your Life

Harold E. Robles

Albert Schweitzer: An Adventurer for Humanity

Kamin Samuel, PhD

Wealth Creation for Coaches: A Workbook to Create a Prosperous Coaching Practice One Small Step at a Time (Kamin Samuel, PhD and Steve Chandler)

Albert Schweitzer

Reverence for Life: The Words of Albert Schweitzer

Patrick O. Smith

ACDF: The Informed Patient: My journey undergoing neck fusion surgery

William Tillier

Abraham H. Maslow: A Comprehensive Bibliography
Personality Development through Positive Disintegration: The Work of Kazimierz Dąbrowski

Margery Williams

The Velveteen Rabbit: or How Toys Become Real

www.MauriceBassett.com